Marina V. Sokolova and Antonio Fernández-Caballero

Decision Making in Complex Systems

T0180126

Intelligent Systems Reference Library, Volume 30

Editors-in-Chief

Prof. Janusz Kacprzyk
Systems Research Institute
Polish Academy of Sciences
ul. Newelska 6
01-447 Warsaw
Poland
E-mail: kacprzyk@ibspan.waw.pl

Prof. Lakhmi C. Jain
University of South Australia
Adelaide
Mawson Lakes Campus
South Australia 5095
Australia
E-mail: Lakhmi.jain@unisa.edu.au

Further volumes of this series can be found on our
homepage: springer.com

Marina V. Sokolova and Antonio Fernández-Caballero

Decision Making in Complex Systems

The DeciMaS Agent-Based Interdisciplinary Framework Approach

 Springer

Dr. Marina V. Sokolova
Albacete Universidad de Castilla-La Mancha
Departamento de Sistemas Informáticos &
Instituto de Investigación en Informática
02071 Albacete
Spain
E-mail: marina@dsi.uclm.es

Prof. Antonio Fernández-Caballero
Albacete Universidad de Castilla-La Mancha
Departamento de Sistemas Informáticos &
Instituto de Investigación en Informática
02071 Albacete
Spain
E-mail: Antonio.Fdez@uclm.es

ISBN 978-3-642-44444-9

ISBN 978-3-642-25544-1 (eBook)

DOI 10.1007/978-3-642-25544-1

Intelligent Systems Reference Library

ISSN 1868-4394

Typeset by Scientific Publishing Services Pvt. Ltd., Chennai, India.

Printed on acid-free paper

9 8 7 6 5 4 3 2 1

springer.com

This book is dedicated to our families and friends, supporting us all around the world.

Preface

From the definition that "a system is a set of elements which interact with each other and the environment", one can note how the number of elements of a system and the number and nature of interactions that occur therein provide the information needed to estimate the degree of difficulty involved to accurately describe it. Clearly, when a system consists of numerous elements, knowledge of the relationships between them is crucial. The mere identification of these elements is not enough to understand its operation, and could lead to form a misconception or insufficient idea of the system as a whole.

Therefore, the study of complex systems attracts the attention of many researchers in all fields. Decision making for complex systems has traditionally been a complicated task that can be complied using an interdisciplinary approach. It has numerous outcomes and can be applied to manifold domains. Complex systems models highlight the dynamics of change. These complex systems are characterized by a high number of component entities and a high degree of interactions. One of the most important features is that it does not involve a central organizing authority, but the various elements that make up the system are self-organized. Some complex systems possess an emergency priority: climate change and sustainable development research, studies of public health, ecosystem habitats, epidemiology, and medicine, among others.

A great number of overlapping approaches that exist nowadays fail to meet the needs of decision makers when managing complex domains. The design of complex systems often requires the integration of a number of artificial intelligence tools and techniques. The problem can be viewed in terms of goals, states, and actions, choosing the best action to move the system toward its desired state or behavior. Agent-based approaches are increasingly being used to model complex systems as they have the advantage of being easy to describe and appear more realistic.

The main objective of this book is to bring together existing methods for decision support systems creation within a coherent framework and to provide an interdisciplinary and flexible methodology for modeling complex and systemic domains and policies. It is obvious that it is impossible to create an overall tool that would play the role of a magic wand to study any complex adaptive problem. But what can be done, and what is absolutely necessary, is a framework for decision support systems design.

This book consists of six chapters. Chapter 1 is dedicated to making a brief introduction into complex system analysis with a particular emphasis on the problem of decision making for such systems. The chapter provides a review on challenges and approaches of modern decision support systems and shows that it is necessary to work out a new agent-based framework for decision support system creation for the case of complex domains. Next, the chapter gives a general overview of the problem at hand, the motivation and some highlights for this study.

Chapter 2 is devoted to related research and introduces an overview of current state-of-the-art in areas of complex system analysis and decision making by means of intelligent tools, including those that are agent-based. It also compares existing agent-oriented methodologies and frameworks for decision support. Furthermore, the chapter makes conclusions about the necessity of a general framework, which would combine characteristics of the former and the latter.

Chapter 3 is a description of the proposed DeciMaS framework. The chapter introduces components of the DeciMaS framework and the way in which they are organized. Design and implementation of the system are discussed as well. The chapter demonstrates how information is transformed into knowledge and illustrates the transformations made throughout each stage of the DeciMaS framework. Furthermore, the information about data mining methods, that are used in the DeciMaS framework, are introduced and depicted in detail.

Chapter 4 provides a case study of an agent-based decision support system developed in an effort to demonstrate the practical aspects of developing systems using the DeciMaS framework. This chapter presents an example of using the DeciMaS framework with the aim of developing a decision support system for environmental impact assessment upon human health. Meta-ontology of the domain of interest, the system itself and its mapping are presented in the chapter. In addition, a multi-agent architecture for a decision support system is shown. The sequence of the steps for the DeciMaS framework design with Prometheus Development Kit and its implementation with Jack Development Environment are presented as well.

Chapter 5 is dedicated to the discussion of the results obtained from the experiment for the case of environmental impact for some selected regions. Data and experiment results of data modeling, simulation, sensitivity analysis, impact assessment and decision generation are discussed.

Chapter 6 contains the conclusions of the proposed work and provides a list of activities for the near future.

Albacete, Spain,
September 2011

Marina V. Sokolova
Antonio Fernández-Caballero

Acknowledgements

Completing a book is truly a marathon event, and we would not have been able to complete this journey without the help and support of countless people over the past four years. Al these anonymous supporters know of our everlasting gratitude.

Marina is especially grateful to all her friends and colleagues from the Kursk State Technical University for their assistance with all kind of help and for their faith in her. During years 2006 to 2008 she enjoyed the aid of a MAEC Scholarship from the Spanish Ministry of Foreign Affairs, which supported her while she started her PhD in the Advanced Informatics Program of the University of Castilla-La Mancha, Albacete, Spain.

We are very grateful to all our friends and colleagues from the Department of Computing Systems at University of Castilla-La Mancha and colleagues from the Albacete Research Institute of Informatics. Let us also say "thank you" to the members of the n&aIS (natural and artificial Interactions Systems) research laboratory for helping us at every time.

Finally, we are forever indebted to our families for their understanding, endless patience and encouragement when it was most required.

Contents

Acronyms

AA   ANN agent
ACL   Agent Communication Language
ADSS   Agent-based decision support system
ANN   Artificial neural network
AOR   Agent-Object-Relationship
AOS   Agent-oriented software
AOSD   Agent-oriented software development
AUML   Agent-based Unified Modeling Language
BOD   Biochemical oxygen demand
BP   Backpropagation
COD   Chemical oxygen demand
CS   Complex system
DM   Data mining
DOA   Domain Ontology agent
BDI   Belief-desire-intention
CM   Committee machine
CMA   Committee Machine agent
CORMAS   COmmon pool Resources and Multi-Agents Systems
CSV   Comma-separated values
DA   Decomposition agent
DBMS   Database management system
DeciMaS   Decision Making in Complex Systems
DGMS   Dialog generation and management system
DPA   Data Preprocessing agent
DSA   Data Smoothing agent
DSS   Decision support system
EA   Evaluation agent
EIA   Environmental impact assessment
EIS   Environmental information system
ES   Expert system
FIPA   Foundation for Intelligent Physical Agents

GA   Genetic algorithm
GAA   Gaps and Artifacts Check agent
GMDH   Group Method of Data Handling
GMDHA   Group Method of Data Handling agent
GOPRR   Graph-Object-Property-Role-Relationship
ICD   International Statistical Classification of Diseases and Related Health
            Problems
ICIDH   International Classification of Functioning and Disability
IDSS   Intelligent decision support system
IDK   INGENIAS Development Kit
IQR   Interquartile range
ISA   Integrated sustainability assessment
JADE   Java Agent DEvelopment framework
JVM   Java Virtual Machine
LCA   Life cycle assessment
LGPL   Lesser General Public License
MAS   Multi-agent system
MASDK   Multi-agent System Development Kit
MBMS   Model-based management system
MiFA   Mining Data Fusion agent
MoFA   Morbidity Data Fusion agent
NA   Normalization agent
OMT   Object Modeling Technique
OO   Object-oriented
PFA   Petroleum Data Fusion agent
PDT   Prometheus Design Tool
RA   Regression agent
RPROP   Resilient propagation
RUP   Rational Unified Process
TFA   Traffic Pollution Fusion agent
UML   Unified Modeling Language
WDFA   Waste Data Fusion agent
WFA   Water Data Fusion agent
WWTP   Waste water treatment plant
XML   Extensible Markup Language
XSLT   XSL Transformations

Chapter 1
Decision Making in Complex Systems

Thoughts and ideas are the source of all wealth,success,
material gain, all great discoveries, inventions and
achievements.
Mark Victor Hansen

Abstract. This chapter introduces the main concepts related to this book. The chapter starts offering a definition of complex systems and introduces some of their principal characteristics. Then the main stages for system analysis are shown. These are system description, system decomposition, study of subsystems, and, finally, integration/aggregation of results. Afterwards, the concepts of decision making and decision support systems are explained. Next, the evolution of decision support systems as well as an introduction into decision making in complex systems are introduced. It is clearly stated that the principal objective of complex system study and analysis is the possibility not only to understand and describe it, but fundamentally to be able to forecast, control and manage it. It is concluded in this chapter that the main motivations for writing this book are twofold. Firstly, there is a real need for the construction of frameworks for the design of decision support systems. In second place, we demonstrate the need for an interdisciplinary approach when facing decision making in complex systems. Lastly, this chapter confirms the usefulness of the multi-agent paradigm for complex systems modeling. A complete comparison between complex systems' and multi-agent systems' characteristics is performed.

1.1 Introduction

Human activity increases constantly, and both the scale and speed of human influence on the natural, social, economic, and other processes has grown significantly. Therefore, now it is impossible not to take it into account as one of the driving forces in the "human - nature - technology" arena. Furthermore, not only must the man-made activity be taken into account, but also the correspondent interactions and feedbacks, and this portfolio of emerging hazards compose the complex systems, which are the object of this research. The science of today has produced significant results in modeling and controlling man-made technical systems. Notwithstanding, effective management of natural complex phenomena often lies beyond our reach.

M.V. Sokolova, A. Fernández-Caballero: Decision Making in Complex Systems, ISRL 30, pp. 1–18.
springerlink.com   © Springer-Verlag Berlin Heidelberg 2012

1.2 Modeling of Natural and Complex Phenomena

Generally speaking, a complex systems (CS) is a composite object that consists of many heterogeneous (and, on many occasions, complex as well) subsystems. It also incorporates emergent features which arise from interactions within the different levels. Such systems behave in non-trivial ways, originated in composite functional internal flows and structures of the CS. As a general rule, researchers face difficulties when trying to model, simulate and control complex systems.

Due to these facts, it would be correct to say that solving the natural complex systems paradigm is one of the crucial issues of modern science. Because of the high complexity of these complex systems, traditional approaches fail in developing theories and formalisms for their analysis. Such a study can only be realized through a cross-sectoral approach, which uses knowledge and theoretical backgrounds from various disciplines, as well as collaborative efforts of research groups and interested institutions.

1.3 Complex Systems: Definition and Principal Characteristics

The majority of real-life problems related to sustainable development and environment can be classified as complex and composite, and, as a result, they possess some particular characteristics. This is why they require interdisciplinary approaches for their study. Complex systems interweave with numerous social, technological, and natural processes, and they have connections with various institutions that impede making mono-solutions and/or taking mono-approaches.

As a rule, according to Levin [116], a "system" is determined to be a "set of interdependent or temporally interacting parts", where parts are, generally, systems themselves, and are composed of other parts in turn. Another definition of a system, given by Rechtin [152] is as follows:

"A system is a construct or collection of different elements that together produce results not obtainable by the elements alone. The elements, or parts, can include people, hardware, software, facilities, policies, and documents; that is, all things required to produce systems-level results. The results include system level qualities, properties, characteristics, functions, behavior and performance. The value added by the system as a whole, beyond that contributed independently by the parts, is primarily created by the relationship among the parts; that is, how they are interconnected".

If any part is being extracted from the system, it loses its particular characteristics (emergency), and converts it into an array of components or assemblies [127]. An effective approach to CS study has to follow the principles of system analysis, when we have to switch over to the abstract view of the system and perform the following flow of tasks:

Table 1.1 Methods for system analysis

Stage	Realization
System description	Expert description, graphical methods, graphs, Petri nets, hierarchies, "AND-OR" and morphological trees, ontologies
System decomposition	Criterion-based methods, heuristic approaches, alternative-based methods, hierarchical methods
Study of subsystems	Problem-solving methods, data mining techniques, knowledge management tools
Integration / Aggregation of results	Decision building tools, (hierarchical selection and composition), data fusion methods

- *Description of a system.* Identification of its main properties and parameters.
- *Study of interconnections* amongst parts of the system, which include informational, physical, dynamical, and temporal interactions, as well as the functionality of the parts within the system.
- *Study of external system interactions* with the environment and with other systems, etc.
- *System decomposition and partitioning.* Decomposition supposes the extraction of series of system parts, and partitioning suggests the extraction of parallel system parts. These methods are based on cluster analysis (iterative process of integration of system elements into groups) or content analysis (system division into parts, based on physical partitioning or function analysis).
- *Study of each subsystem or system part*, using optimal corresponding tools (multidisciplinary approaches, problem solving methods, expert advice, knowledge discovery, and so on).
- *Integration of results* received from the previous stage, and obtaining of a pooled knowledge about the system. Synthesis of knowledge and composition of a complete model of the system. It can include formal methods for design, multicriteria methods for optimization, decision-based and hierarchical design, artificial intelligence approaches, case-based reasoning, and others (for example, hybrid methods [183], [142]).

In Table 1.1 we offer some possible approaches and tools that can be used at each stage and are not limited by the mentioned methods. To view in more detail the stages - provided in the first column of Table 1.1 - and the methods - offered in the second column-, we name the more frequently used ones. The first stage is "System description", which serves as a background for future system analysis. On this stage the knowledge, the rules, and the database are created. A number of varied methods can be applied here:

1. Expert description, essential in this case, can be combined with graphical methods of system representation (data flow diagrams, decision tables, decision trees, and so on).
2. Graphs in form of flow graph representations, potential graph representations, resistance graph representations, line graph representations.
3. Petri nets.
4. Taxonomies, vocabularies, and various kinds of hierarchies.
5. "AND-OR" trees, morphological trees and their variants.
6. Ontologies, which unify necessary expert and technical information about the domains of interest and databases.

"System decomposition" is the second stage. It is necessary for studying the system components. Among the possible solution methods we can name criterion-based methods. The third stage, "Study of subsystems", implies a wide usage of problem-solving methods, which are designed to describe knowledge and reasoning methods used to complete a task. Then, in this stage there is a great variety of data mining techniques to solve the following tasks:

1. Classification
2. Regression
3. Attribute importance
4. Association
5. Clustering

The specific methods, which are used for this list of tasks, include statistics, artificial intelligence, decision trees, fuzzy logic, etc. Of course, there are a number of novel methods, modifications and hybridization of existing tools for data mining that appear permanently, and are successfully used at this stage. At the fourth stage, "Integration / Aggregation of results", sets of methods dedicated to integration and composition are used:

1. Methods for data integration
2. Multi-criteria evaluation
3. Decision making, including group and distributed decision making
4. Evolutionary methods
5. Artificial intelligence methods, and so on.

To be able to analyze a system in practice, a researcher has to accept some assumptions about the detailed elaboration level of the system. This means that the researcher has to choose those entities considered to be the elemental ones for research. As a rule, this task is assumed by specialists. Generally, there are various types of specialists, who collaborate on the same case study. Levin [116] enumerates the traditional classes of specialists:

1. Novice or beginner
2. Professional or competent
3. Expert

Notice that Levin cites the classification of Denning [39], who suggests a seven-level specification:

1. Novice or beginner
2. Advanced beginner or rookie
3. Professional or competent
4. Proficient professional or star
5. Expert or virtuoso
6. Master
7. Legend

The role played by specialists, especially of the higher levels, can not be underestimated, as they determine the informational support and ontological basis of the research. Of course, the qualitative and quantitative outcomes of research have to be evaluated. In the case of complex systems we may need to use different types of criteria, often hybrid ones, and composite scales as mono-scaled viewpoints have proved themselves to be incorrect and inappropriate for CS solutions [160]. Usually, in such cases they have to be provided as local measurements for each component of the abstract system model, and general evaluation criteria for the system can be received as a fused hybrid estimation.

Complex adaptive systems are characterized by self-organization, adaptation, heterogeneity across scales, and distributed control. Though these systems are ill-determined, ambiguous and uncertain, they have commonalities which may help working out a general approach to their study. Complex systems, or systems of systems, are characterized by a high complexity and a great number of interacting components. Here, difficulties have already appeared during the creation of an abstract model of a complex system, because of the great number of decisions to be made regarding its design. Non-traditional tools from different domains can be highly effective in the case of a complex system, providing novel ways to generate decisions and to find solutions. Complex systems cover several dimensions, including economical, ecological, and social sub-systems, which may have any level (positive or negative) of collaboration and acceptance between them.

1.4 Decision Making and Decision Support Systems

1.4.1 Decision Making

Decision making is the final goal of a decision support system (DSS), and, for complex domains, it is multi-sectorial by nature. In practice, decisions can be taken by single authorities or by a group of responsible decision makers. Nowadays, decision makers use DSS which serve as an informational background, enabling real-time simulation and further decision generation. The different scales of a decision making process, for instance in a human health domain, may include a wide range of organizations at different levels: international, national and local. Informational support, received with DSS, helps to mobilize and allocate resources, to set priorities and to share successful patterns and strategies.

1.4.2   The Evolution of Decision Support Systems

The idea of creating a symbiotic human-computer system, with the aim of incrementing accessible knowledge for decision making in complex problems, appeared in the mid-sixties, and was first applied to managerial and business domains. During this initial period of CS, the development of principal components and fundamental characteristics of such systems started. In 1971, Gorry and Scott-Morton [67] suggested using the term "decision support system" for supporting information systems that generate semi-structured and unstructured decisions. Later, the initial domains of DSS applications (in financial and business areas) were widened, and the concept of DSS was spread to manifold spheres and fields of human activities, extending to technical as well as to complex and easily-determined domains (environmental, medical, social issues, and so on). Over time, diverse DSS, which embrace a number of models for use such as preprocessing, optimization, hybrid, and simulation models, have appeared.

Nowadays, there are many definitions of what a DSS is. For example, a DSS can be designed as a specific class of computerized information system that supports business and organizational decision-making activities. A properly designed DSS is an interactive software-based system intended to help decision makers in compiling useful information from raw data, documents, personal knowledge, and/or business models to identify and solve problems and to make decisions. Power [149] gives a full description of this process in his review of DSS history and evolution. In agreement with this review, Bonczek and collaborators [15] define a DSS as a computer-based system consisting of three interacting components: a language system, a knowledge system, and a problem-processing system. This definition covers both old and new DSS designs, as the problem processing system could be a model-based system, an expert system, an agent-based system, or some other system providing problem manipulation capabilities. Keen [105] applies the term DSS to situations where a "final" system can be developed only through an adaptive process of learning and evolution. Thus, the author defines a DSS as the product of a developmental process involving the builder, the user and the DSS itself to evolve into a combined system.

Sprague and Carlson [188] identify three fundamental components of a DSS:

1. Database management system (DBMS)
2. Model-based management system (MBMS)
3. Dialog generation and management system (DGMS)

Moreover, Haag and colleagues [71] describe these three components in more detail. The Data Management Component stores information that can be further subdivided into that derived from an organization's traditional data repositories, from external sources, or from the personal insights and experiences of individual users. The Model Management Component handles representations of events, facts, or situations, using various kinds of models, two examples being optimization models and goal-seeking models. And the User Interface Management Component is, of course, the component that allows a user to interact with the system. According to

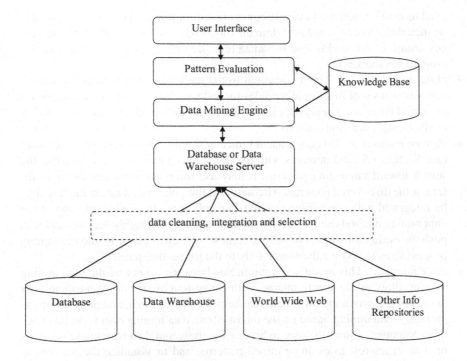

Fig. 1.1 Architecture of a typical data mining system, adapted from [73]

Power [149], academicians and practitioners have discussed building DSS in terms of four major components:

1. User interface
2. Database
3. Modeling and analytical tools
4. DSS architecture and network

Thus, the architecture of a typical data mining system may have the following major components (as shown in Fig. 1.1) [73]:

- Database, data warehouse, World Wide Web, or any other *information repository*. This is, it may be one or a set of databases, data warehouses, spreadsheets, or other kinds of information repositories. Data cleaning and data integration techniques may be performed on the data.
- *Database or data warehouse server*. The database or data warehouse server is responsible for fetching the relevant data based on the user's data mining request.
- *Knowledge base*. This is the domain knowledge that is used to guide the search or to evaluate the interestingness of the resulting patterns. Such knowledge can include concept hierarchies, used to organize attributes or attribute values into different levels of abstraction. Knowledge such as user beliefs, which can be

used to assess a pattern's interestingness based on its unexpectedness, may also be included. Other examples of domain knowledge are additional interestingness constraints or thresholds, and metadata (e.g., describing data from multiple heterogeneous sources).

- *Data mining engine.* This is essential to the data mining system and it ideally consists of a set of functional modules for tasks such as characterization, association and correlation analysis, classification, prediction, cluster analysis, outlier analysis, and evolution analysis.
- *Pattern evaluation.* This component typically employs interestingness measures (see Section 1.5) and interacts with the data mining modules so as to focus the search toward interesting patterns. It may use interestingness thresholds to filter out the discovered patterns. Alternatively, the pattern evaluation module may be integrated with the mining module, depending on the implementation of the data mining method used. For efficient data mining, it is highly recommended to push the evaluation of pattern interestingness as deep as possible into the mining process so as to confine the search only to the interesting patterns.
- *User interface.* This module communicates between users and the data mining system, allowing the user to interact with the system by specifying a data mining query or task, providing information to help focus the search, and performing exploratory data mining based on the intermediate data mining results. In addition, this component allows the user to browse database and data warehouse schemes or data structures, to evaluate mined patterns, and to visualize the patterns in different forms.

The most important point is that, despite a great diversity of DSS realizations, the set of general aims remains the same and includes the following items:

- Helping to advance in knowledge and to understand the situation or object under study.
- Assisting in making planning decisions.
- Organizing multi-access interactive computing.
- Calculating and estimating the consequences of possible actions on the basis of simulation models.
- Imitating human reasoning and using it for simulation.
- Providing guidelines for action by generating sets of optimal solutions.
- Suggesting models or constructing solutions that may lead to specific decisions.
- Improving decision making.
- Integrating various decision making approaches.
- Providing possibilities for distributed and shared decision processes.

In order to add more functionality and to make the DSS more flexible and autonomous, the MBMS can be replaced with Expert Systems (ES) or other intelligent decision making blocks. Here, artificial intelligence, computational agents and fuzzy-based methods are applied. The advantages of using intelligent components with DSS are increased timeliness in making decisions, improved consistency in decisions, improved explanations and justifications for specific recommendations, improved management of uncertainty, and formalization of organizational knowledge.

The most useful of these advantages is improved explanations and justifications, which is an extremely useful feature, particularly in the study of complex systems, where it helps if the real expert can validate the machine reasoning process.

Improved decision making represents an optimal background for practical DSS creation. However, it must also contain the four subsystems that Power describes, that is, a database attached (data subsystem), a mechanism for processing the data and obtaining knowledge from there (expert subsystem), which could be represented by models, rules, agent-based subsystems or some other intelligent techniques and must be capable of receiving and acting on requests from users (user interface subsystem). These components have to be placed together in accordance with the DSS ontology (the DSS architecture and network).

1.4.3 Decision Making in Complex Systems

The principal objective of CS study and analysis is the possibility not only to understand and describe, but fundamentally to be able to forecast, control and manage it. It is clear that in case of CS we cannot attempt to employ ideas of rigid "command - execution" style management. On the contrary, we can only rely on flexible preemptive/anticipated correction, which would be harmonized with the nature and dynamics of the respective CS. The degree of control over different components of a CS varies. Indeed, the technical or managerial subsystems can be controlled, but the independent components of the CS generate their own uncontrolled decisions. Here, the classical decision making process converts into a shared and constant collaboration between specialists and in a support tool. For example, there may be a system for support in decision making to preview and to gently correct CS states and tendencies with respect to its self-organization, emergent behavior, and adaptation over time.

In addition, the DSS design and development process has to be based on an integrated approach, which allows the interconnection of all the tools of the decision support techniques portfolio with the organization's transactional systems and the external and internal data flows. Generally, basic components for making decisions are found in any DSS. These are:

1. Knowledge base
2. Models base
3. Rules base
4. Possible situation bank
5. Learning flow/block/module/feedback

The structure of a DSS has to satisfy the requirements imposed by specialists and the characteristics and restrictions (the limits as well as the uncertainty) of the application domain. In Fig. 1.2 there is a general work flow of a decision making process embodied in a DSS. The traditional "decision making" work flow includes the preparatory period, the development of decision and, finally, the decision making itself and its realization [177].

Phase 1. Decision preparation

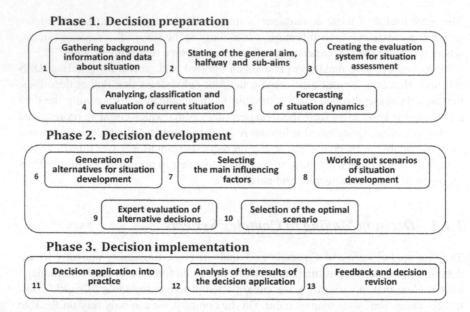

Phase 2. Decision development

Phase 3. Decision implementation

Fig. 1.2 The general work flow of a decision making process

The preparatory period, "Phase 1. Decision preparation", consists of several intermediate stages. The initial one is acquisition of background knowledge about input and output information, as well as about the possible and the desired situations (stage one, "Gathering background information and data about situation"). This knowledge forms the data base, the knowledge base and the rule base. Next, on stage two, "Stating of the general aim, halfway and sub-aims", the hierarchy of aims, where general objectives, halfway aims and sub-aims are stated, is created. The next important stage three is the elaboration of the evaluation system, that is "Creating the evaluation system for situation assessment". Having solved these three stages, we move to system analysis and diagnostic (stage four, "Analyzing, classification and evaluation of current situation"), and, then, to the final stage of the preparatory period, namely the forecasting module (stage five, "Forecasting of situation dynamics").

The development of decision or decision making, "Phase 2. Decision development", covers the stages ranging from six to ten, referring to the numbering at Fig. 1.2. The stage number 6, "Generation of alternatives for situation development", suggests the use of a portfolio of tools and knowledge obtained during the previous period, and simulation of possible outcomes of situation behavior. Then, on the next step 7, "Selecting the main influencing factors", the factors of interest are selected in order to check their influence on the studied processes. With respect to the factors and using a set of alternatives (at stages six and seven), scenarios of situation development are worked out (stage eight, "Working out scenarios of situation development") and then (stage nine, "Expert evaluation of alternative decisions") evaluated by groups of experts. The final alternative - or a ranged set of alternatives together with the correspondent constructed solutions and controlling actions - is received as a result of their decision and is formulated on stage ten, "Selection of the optimal solution".

The third period, "Phase 3. Decision implementation", is related to decision implementation into practice. The important point here is the result of the decision application, "Decision application in practice" (stage 11), its impact on the problem in question, "Analysis of the results of the decision application" (stage 12), and the received feedback, "Feedback and decision revision" (stage 13).

1.5 Composite Decisions

Because of the specific characteristics of each application domain, the term "decision" can be denoted in different ways. Hence, in general terms, "decision" is an administrative action or a solution of a task. It is integrated in the interaction circle "human - environment" and is a main objective of the decision making process. Coming to a decision in the case of an organization means finding a compromise, which has to comply with its mission, tactical and strategic goals for artificial organizations, and to accomplish and facilitate reaching the sustainable optimum and survival conditions for natural organizations. In case of mixed organizational structures decision making processes should be flexible enough to satisfy requirements put on by complexity of internal and external interactions and to achieve manifold goals with respect to the nature of each component. When looking at the complexity of the issue under scrutiny, it seems reasonable that a decision making process should intrinsically involve the whole portfolio of possible tools. Along the same line is the necessity to provide group and distributed decision making, and to facilitate collaborative efforts.

In accordance with Fig. 1.3, a decision can be viewed as an intersection of the spaces of possible decisions, possible alternatives and selection criteria. The complexity increases in cases in which all these spaces have a composed organization. In the simplest case, possible alternatives are independent, but they can be grouped into clusters, or form hierarchies; decisions can consist of the best optimal alternative, but they can also be formed as a result of a combination of alternatives (linear, non-linear, parallel, and so on), and their subsets and stratifications; criteria can be both independent or dependent, and, commonly, hierarchically organized.

In the scientific research there are several approaches to the decision making process; for example, decision tables, decision trees and flows. Composite decision making is hierarchical, and various decision making methods and tools are used at different levels. The composite decision is not final either, as it is corrected, accepted or rejected by a specialist. The general schema of composite decision making is shown in Fig. 1.4.

From a practical point of view, a decision making process produces a number of problems, mostly related to the design and implementation periods of the decision making work flow. These problems include excessive or scarce information, multiple heterogeneity of information sources and contents, and lacks of data quality.

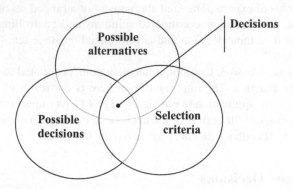

Fig. 1.3 Decision as a result of the intersection between possible decisions, alternatives, with respect to selection criteria

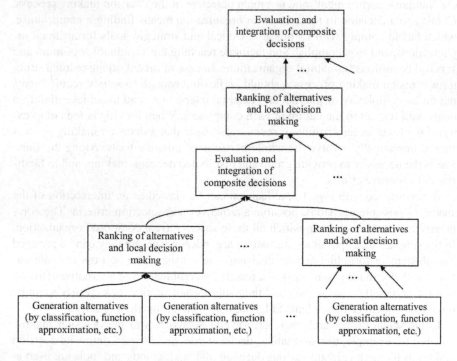

Fig. 1.4 Bottom-up hierarchical decision making, adapted from [116]

Dealing with data mining, a number of methods can be put into practice, but it is hard to select the optimal modeling strategy for each component of the system. It makes intelligent data analysis of special importance here.

1.6 Motivation

1.6.1 Frameworks for the Design of Decision Support Systems

It is obvious that it is impossible to create an overall tool that would play the role of a "magic wand" to study any complex adaptive problem. But what can be done, and what is absolutely necessary, is creating a framework for DSS design. In general terms, a framework facilitates the development of an information system, because it permits organization of necessary goals and works into workflows. In our case the information system in the concept definition given above represents a CS or a natural phenomenon. By following the stages determined in a framework, developers and programmers will have more possibilities to specialize in the domain of interest, and to meet the software requirements for the problem in term, thereby reducing the overall development time. Unmistakably, to be effective, the framework should, on the one hand, be general enough to make the system applicable for various domains and, on the other hand, adaptable for specific situations and problem areas. To comply with all these requirements, we should follow the natural flows of the information systems life cycle. As widely described in scientific reviews, the general steps of informational system creation are the following ones [177]:

- Domain Analysis - is related to the analysis of the project idea, the problem definition, the extraction of aims/objectives, the creation of goal trees, and the production of the sequence of tasks and subtasks to be solved. This stage also implies the creation of the domain ontology, which covers the problem area, the set of relations between the concepts and the rules to acquire new knowledge. The work of domain area experts is required at this stage.
- Software Elements Analysis - this stage also deals with the creation of private ontologies, but now the ontologies are created for the system and its elements. The sets of goals and tasks are related to the sets of system functions (roles), required resources (commonly in form of informational files), interactions, and so on.
- Specification - is the written description of the previous stages, which results in the creation of the system meta-ontology.
- Software Architecture - implies the abstract representation of the system to meet the requirements. The software architecture includes the interfaces for computer-user communication.
- Implementation (coding) - the iterative process of program creation.
- Testing - program testing under normal and critical conditions.
- Deployment and Maintenance - program application and support whilst the software is in use. Sometimes training classes are added to the software product.
- End of Maintenance - is the final stage of the software life-cycle.

Complex systems require multidimensional hierarchical solutions, and, in agreement with our arguments about the nature of complex natural phenomena and their study (see Chapter 1.3), we offer an interdisciplinary approach, which combines theories and formalisms from traditional disciplines in order to focus the integrated efforts from various viewpoints on the problem.

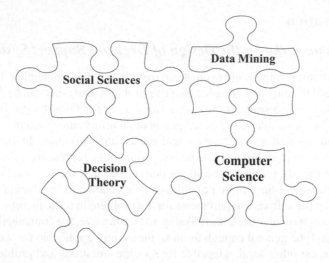

Fig. 1.5 The systemic puzzle for complex systems study

1.6.2   The Need for an Interdisciplinary Approach

One of the principal lessons drawn from scientific issues related to CS analysis is
that any mono-disciplinary based approach has failed in analyzing complex environmental and social phenomena. In accordance with the main principals of the system
approach, we propose an integrated view, based on systems analysis, concentrating
on the study of the interaction mechanisms among the different subsystems. Here
the term "integrated" stands for a broad interdisciplinary approach, where a wide
range of manifold approaches from "human", "technological" and "environmental"
systems such as Social Science, Data Mining, Decision Theory and Computer Science are reused (see Fig.1.5) [131], [46].

Taking into account the complexity of the interaction between the components
of complex systems, which are self-organized adaptive systems themselves, we will
need a cross-sectoral approach to ensure that various components of this system are
not studied in isolation. In time to search for the appropriate analysis instruments,
we have to clarify what kind of combination between existing approaches have to
be given preference. In accordance with [171], a *multidisciplinary approach* can be
determined as "a term referring to the philosophy of converging multiple specialties
and/or technologies to establish a diagnosis...", and an *interdisciplinary approach* is
"what results from the melding of two or more disciplines to create a new (interdisciplinary) science". The multidisciplinary approach has been widely used in research
practice (and can be compared to collective decision making or to hybridization),
while the interdisciplinary approach supposes a new way of organization, processing and managing information. Being applied to a complex system, the approach
includes the following main works:

1. The complex system, for example "human" (H), "technology" (T) and "environment" (E) system is decomposed into components of the "second" layer: H, T and E subsystems or, if necessary, the process of decomposition is repeated.
2. Then, each of the subsystems is studied by means of the "proper" techniques, belonging to the respective discipline, or/and of hybrid methods.
3. The managerial process of decision making is realized, with feedback and generation of possible solutions.

1.6.3 Multi-agent Paradigm for Complex Systems Modeling

Agents and multi-agent systems (MAS) are actively used for problem solving and have recommended themselves as a reliable and powerful technique [200], [206], [84], [56], [181], [55]. The term "agent" has many definitions, and is commonly determined as "an entity that can observe and act upon an environment and directs its activity towards achieving goals". Commonly, an agent has some characteristics, which are [184], [121]:

- reactivity (an agent responds in a timely fashion to changes in the environment);
- autonomy (an agent exercises control over its own actions);
- goal-orientation (an agent does not simply act in response to the environment, but attempts to achieve its goals);
- learning (an agent changes its behavior due to its previous experience);
- reasoning (the ability to analyze and make decisions);
- communication (an agent communicates with other agents, including external entities); and,
- mobility (an agent is able to transport itself from one machine to another).

In practice, agents are often included into multi-agents systems, which can be determined as a decentralized community of intelligent task solving entities (agents), oriented towards some problem [51], [207], [101], [54].

A multi-agent system itself is a clear example of a complex system, because the former shares all the properties of the latter. Let us see how MAS meets these characteristics. Table 1.2 shows some comparison results between a complex system and a multi-agent system based on the analysis of the properties of the former and the correspondent characteristics of the latter. Table 1.2 is organized as follows: the first column contains a systemic property of a complex system, the second and third ones explain how this property is applied/discovered for a CS and for a MAS, repectively.

"Emergency" is the fundamental characteristic of complexity and both paradigms (CS and MAS) share it. In case of CS and MAS it leads to appearance of new properties. "Internal interactions" are represented by message exchange that is collected in protocols in case of MAS; and in case of CS, the internal interaction of its components can be described by various means (physical, informational, functional interactions, etc.) but is always present. Every CS can be decomposed into sub-systems; in case of MAS, it consists of agents/agent teams. The fact that both of them do not

Table 1.2 Comparison between CS and MAS characteristics

Systemic property	Complex system	Multi-agent system
Emergency	The mutual relationships among the components of the system result in spontaneous emergence of new properties and behavior patterns.	The interaction of agents within MAS leads to collective interactions and new patterns emerge from different agent interactions.
Internal interactions	The external interactions among the components of a CS are stronger than interactions with any component of the environment.	The agents within a MAS continuously interact with each other and the response to an external entity is the response of the overall MAS.
System consists of components	A system is a construct or collection of different elements; each of which can represent a system itself.	A MAS is a community of autonomous (or semi-autonomous) agents or agent teams.
Decentralization	The components of a system represent heterogeneous domains of different nature and with self-organization and control.	No agent has a full global view of the system, or the system is too complex for an agent to make practical use of such knowledge.
Composite decision making	The decision is a result of a hierarchical selection and composition of local decisions.	The decision is a result of "pooled" decisions made by the agents.

execute overall centralized control is reflected in the "decentralization" property. The final decision for both CS and MAS is a result of a weighted collective choice (see Fig. 1.3).

Agents have abilities to represent both the dynamic behavior by using sets of rules, and the emergent behavior interacting with other agents [160]. Multi-agent systems manifest self-organization and complex behavior even when the individual strategies of all their agents are simple. The agents share knowledge using any agreed language within the constraints of the system's communication protocol. Some examples of agent communication languages are the Knowledge Query Manipulation Language (KQML) and the Foundation for Intelligent Physical Agents Agent Communication Language (FIPA ACL) [48]. There is a need of mechanisms for advertising, finding, fusing, using, presenting, managing, and updating agent services and information. Most MAS use facilitator agents to help find agents to which other agents surrender their autonomy in exchange for the facilitator's services. Facilitators can coordinate the agents' activities and satisfy requests on behalf of their subordinated agents.

MAS can be classified in accordance to several classifiers. Firstly, there are closed and open MAS. The former contains well-behaved agents designed to cooperate together easily towards a global common goal. A MAS, related to the latter, can contain agents that are not designed to cooperate and coordinate, but to assist the agents in working together. The most common kind of these mechanisms is for negotiations and auctions. Weiss [200] provides another interesting classification: MAS classified by the level of autonomy, of organizational type, and architecture. Depending on the level of autonomy and self-orientation of every agent, a MAS can vary from distributed and "independent" to supervised systems of "organizational" type, in which every agent knows the order and turn of its execution.

Remember that one of the most important principles of CS is that they can not be studied from a mono-discipline viewpoint. It is necessary to provide a complex hybrid application of methods and techniques from various disciplines. Using agents seems to be an optimal solution in this case. Which are the qualities of MAS that enable us to put them forward as an optimal paradigm? First, a MAS is a community of different agents (different by their nature, by their way of functioning, by their level of autonomy, and so on) built together. In turn, each agent possesses some proper characteristics: autonomy, believes, plans, or self-control. It is capable of reasoning, making decisions, and acting within its sphere of competence. Agents utilize manifold methods. Actually, a MAS helps to create cross-disciplinary approaches for data processing, and, hence, for CS study. Second, an agent may include nontraditional instruments to bear from different domains. A role played by an agent depends on the system (or subsystem) functions and aims. There are no restrictions or limitations put on the knowledge and rule bases used by each agent. During the execution it acts in the following way (for Belief-Desire-Intention (BDI) agents). The way in that an agent handles tasks, satisfies all the main principles of the system approach, which has the following main steps, in accordance with [200]:

- Task decomposition. Decomposing complex tasks to subtasks and forming the task execution flow.
- Task allocation. Assigning tasks to certain agents.
- Task accomplishment. The agents accomplish their tasks, which may include recursive task decomposition and work delegation to free appropriate agents.
- Result synthesis. The backward synthesis of results. The final tasks solvers pass the results to the original agents As a result, a common composite decision is generated.

In third place, MAS ensure dealing with multiple complexity:

1. They work with geographically distributed system components.
2. They manipulate huge amounts of information about concepts, and correspondent operations and calculations.
3. They cover significant parts of an application domain or may run across various domains.
4. They work with multi-component systems.

And, lastly, one of the principal MAS characteristics is that they have an open architecture and do not have centralized control. To sum up, the motivation of this work is to bring together existing methods for decision support systems creation within a more coherent approach and to provide an interdisciplinary flexible methodology for complex, systemic domains and policies.

Chapter 2
A Review on Frameworks for Decision Support Systems

There are two primary choices in life: to accept conditions as they exist, or accept the responsibility for changing them.
Denis Waitley

Abstract. This chapter is devoted to the research related to frameworks for the design of decision support systems. It introduces an overview of current research in areas of complex system analysis and decision making by means of intelligent tools, including those based on the agency paradigm. It also compares existing agent-oriented methodologies and frameworks for decision support. The review includes a series of tables that describe research papers dedicated to decision support systems in complex local and multi-functional systems, as well as methodologies of specific environmental information systems development. Existing agent-based proposals for decision support systems are categorized into frameworks for multi-agent systems planning and software tools for multi-agent systems design and implementation. A broad comparison between agent-based frameworks is offered. Furthermore, the chapter makes conclusions about the necessity of a general framework, which would combine characteristics of the former and the latter.

2.1 Introduction

The review of the current state of the art in the area of complex systems modeling, agent-oriented methodologies, and decision making approaches is the first step before thinking of any new approach in constructing a framework for the design of decision support systems. We are aware that it is not possible to create a unified methodology for the design of decision support systems for complex domains. This goal is extremely difficult to achieve because we have to take into account the similarities shared by diverse complex systems (CS) without losing their specific features. However there are many solutions and multi-function tools, decision support systems perform better results when oriented to limited and determined domains. The problem lies in the existence of a great number of overlapping approaches and methodologies which demonstrate successful results, but nevertheless fail to meet the needs of decision makers for an integrated methodology which supports decision making in complex domains.

M.V. Sokolova, A. Fernández-Caballero: Decision Making in Complex Systems, ISRL 30, pp. 19–45.
springerlink.com   © Springer-Verlag Berlin Heidelberg 2012

However two solutions may be proposed: first, bring together existing methods for decision support systems creation within a more coherent system; second, provide an interdisciplinary flexible methodology for complex, systemic domains and policies.

2.2 Decision Support Systems in Academy and Research

Decision support systems (DSS) have significantly evolved and turned into essential tools. As it is noted in [27], modern institutions and corporations tend to become more widespread and to lose rigidness of their organizational structures. On the other hand, the larger number of senior and medium executives consult with DSS and have daily hands-on experience with them. Thus, DSS has become a crucial part of organizations and not just stand-alone applications [27], [40], [144], [154], [169], and [47]. As it is reported in [27], [105], agent technology has emerged as a promising solution to meet the requirements in order to provide decision support for complex domains. As a rule, agents within the decision support systems must carry out the following functions:

- To search for valuable information and to retrieve it.
- To multiple scale in case of heterogeneous sources of income and outcome information.
- To provide intercommunication between the system and the external information sources (e.g. sensors, remote equipment).
- To check input data for consistency and to preprocess it.
- To execute data mining procedures.
- To experiment with different alternatives and to make recommendations.
- To organize human-computer interactions.

Modern "decision support systems" and "expert systems" are commonly based on intelligent agents, and the concepts of DSS as well as those of ES have also been modified [113], [122], [134]. For example, medicine is a traditional field of DSS application, and some recent academic reports deal with examples of novel usage of agent-based DSS for home and hospital care, pre-hospital emergency care and health monitoring and surveillance [2]. One such application is a distributed DSS for brain tumor diagnosis and prognosis (*HealthAgents*) [62], [2], where agents use both data mining methods and decision making techniques. An application of remote control of patients via clinical DSS is created on the basis of multi-agent methodology, and results in the creation of the SAPHIRE multi-agent system. There are many reasons to apply the agent paradigm: the necessity to interact in heterogeneous distributed environments, the need to provide instantaneous communication of autonomous components in a reactive manner, and the possibility to dynamically create and eliminate agents. It this application, the agents provide all the vital functions of the system.

In [4] the outcomes of the construction and usage of an agent-based environmental monitoring system are presented. It is aimed to provide measurements of meteorological information and air pollution, to analyze them and to generate alarm signals. The system is created by means of the intelligent platform "Agent Academy". The system has a three-leveled organizational structure where data preprocessing, its manipulation and distribution are carried out. The necessary steps for data transformation are executed by the following types of intelligent agents: Diagnosis agents, Alarm agents, Database agents and Distribution agents. In another article, the authors report about the application of the agent paradigm for the evaluation of socially-oriented advertising campaigns aimed to affect consumers' behavior [5]. The authors create social communication models to simulate a public response to mass-media influence and introduce a social grid populated with autonomous consumer agents. This grid is also used to evaluate possible outcomes of the public campaign on water demand control.

Another situation assessment is carried out by an agent-based system created with the MASDK tool [63]. The authors present their approach in situation assessment and explain its possible application in different problem areas. Another approach to complex situation assessment has been presented in [124]. The authors have accepted the JDL (Joint Directors of Laboratories) model as a basis for situation awareness. Their new approach to situation assessment learning is described and the structure of the MAS is presented as well. The agents act within the four levels of the hierarchical JDL-model. In another reference [196], the authors present the framework of a decision support system for water management in the Mediterranean islands, coupling a multi-agent system with a geographic information system. The platform developed makes it possible for users to better understand the current operation of the system, to apprehend the evolution of the situation, and to simulate different scenarios according to the selected water policies and the climatic changes hypothesis. Also recently, the development and experimental evaluation of an Internet-enabled multi-agent prototype called AgentStra [117], for developing marketing strategies, competitive strategies and associated e-commerce strategies has been introduced.

On the other hand, specialists working with environmental sciences and public health store huge volumes of relevant information about pollutants and human health. Continuous processing and maintenance of the information requires substantial efforts from the practitioners and professionals, not only while handling and storing data, but also when interpreting it. Actually, it seems very hard to handle all the data without using data mining (DM) methods, which can autonomously dig out all the valuable knowledge that is embedded in a database without human supervision, providing a full life-cycle support of data analysis. Techniques such as clustering, classification, logical and association rule-based reasoning, and other methods are highly demanded for comprehensive environmental data analysis. For instance, DM techniques for knowledge discovery and early diagnostics are utilized for early intervention in developmentally-delayed children [32].

Reference [33] outlines the MAS "Instrumented City Data Base Analyst", which is aimed to reveal correlations between human health and environmental stress factors (traffic activity, meteorological data, noise monitoring information and health statistics) by using a wide range of DM methods, including regression analysis, neural networks, ANOVA and others. The architecture of the system counts a number of modules placed within four levels. The multi-agent structure includes specific modeling agents, which create models for the environmental stress factors, and are then harmonized by the model co-ordination agent. The "Data Abstractor" is an agent that gets information from sensors, fuses it and preprocesses it. Interaction with humans is provided by the "Reception agent". Reference [49] provides a survey of intelligent-based system for decision support which supports clinical management and research. The author gives a brief introduction into DSS and agent-based DSS, and gives an example of a Neonatal Intensive Care Unit system. In the high-level diagram of the proposal there is the solution manager service, which represents an Intelligent DSS (IDSS). The Analytical Processor of the IDSS joins different types of agents: functional, processing, human and sub-agents, and this processor is aimed to detect trends and patterns in the data of interest.

A group of researchers [156] have presented a number of works dedicated to decision making support in a waste water treatment plant (WWTP). In order to provide an intelligent model and control for the WWTP, the authors use eight sorts of autonomous agents which support the decision-making process. The agents, which interact with the WWTP and the users, are the Monitoring Agent and the Actuation Agent, respectively. The Monitoring Agent captures some data from WWTP through the system of sensors, supplies retrospective information, displays it and provides a basic alarm system. The Actuation Agent analyzes some WWTP parameters, suggests orders, processes alarms and reacts in order to solve them. And, lastly, there are User Agents, which are responsible for the organization of the human-user interface, accept orders and transform them into goals for the multi-agent system. It also receives and displays information and recommendations, and supplies justifications to certain actions. The Monitoring Agent, Actuation Agent and User Agents interact with the "nucleus" of the multi-agent system through the communication system, and then, data is analyzed and transformed respectively by the Modeling Agent, the Predictive Agent, and the Numerical Control Agent. Then it is tested and used by the Expert Reasoning Agent and the Experience Reasoning Agent. The Modeling Agent accepts a selection of simulation and control models, enables and disables the simulation and automatic control of the WWTP. As it incorporates a number of modeling methods, it applies them to simulate specific processes of WWTP and suggests control actions. The Predictive Agent is dedicated exclusively to task prediction, interpreting the results of simulations and selecting the best models. The Numerical Control Agent determines the best control algorithm among the ones available from the Modeling agent. The Expert Reasoning Agent contains the expert knowledge and serves as a rule-based data base, which gives permission to receive any new knowledge. The Experience Reasoning Agent learns new experiences provided by an expert or another source of knowledge whenever a new problem is solved.

Ceccaroni et al. [29] continue to develop the idea and implementation of the DSS noted above and offer an environmental decision support system related with a domain ontology. The authors aim to check if the incorporation of an ontology and DSS, based on rule-based reasoning and case-based reasoning, can improve the generated decisions. The article describes the OntoWEDSS architecture, which is derived from the model of the DSS for WWTP. It has an ontology embedded and a reasoning module, which serves to facilitate knowledge sharing and reuse. The reasoning module is modified, and contains rule-based, case-based and ontology-based reasoning. In the publication dedicated to working out an interdisciplinary approach to solving conflicts in water domain [133], the authors bring together the problem of managing a complex system. The problem consists of a complex application domain (water resources)(A) as well as a wide range of decision makers, experts and other personnel (B). In this work the authors attempt to deal with this composite system and discuss the creation of a conceptual framework, which would be able to solve possible conflicts in system A and simultaneously solve problems in the application domain. The framework presented in [160] is based on the application of the agency paradigm to the Integrated Sustainability Assessment (ISA) Cycle. The ambition of ISA is to provide an international scientific society with a general framework, which would include a variety of assessment tools and methods. Then, the author proposes a two-track strategy: intent to use the current portfolio of ISA tools as efficiently and effectively as possible, and to contribute to the generation of new ISA tools.

Another work [97] presents a study dedicated to application of an integrated assessment approach to catchment management. In [115] a nodal network-based approach to the case of water resource allocation and management is applied. The proposed framework uses a nodal network structure which changes its form depending on the type of decision being made. The framework produces scenarios in the form of "what if" questions related to policy and management of the case of study. The framework presented in [97] is centered on the need for improved techniques of uncertainty and sensitivity analysis that can be taken as a measure of confidence for ranking decisions and making a choice. An approach for the integrated assessment of both technical and valuation uncertainties during decision making, based on Life Cycle Assessment (LCA) has also been presented [7]. The approach is built on three conditions: placing appropriate bounds on particular aspects, non-overlapping alternatives, and conducting a sensitivity analysis for valuating uncertainties. Also, a multi-layer tool for socio-environmental simulation has been proposed [194]. The authors have developed a hybrid system, comprised by CORMAS (COmmon pool Resources and Multi-Agents Systems) and a language Q. CORMAS is a system, based on the multi-agent paradigm, which describes the interactions between the natural environment and humans. Q is a language for describing complex interaction scenarios among agents. The simulator has a two-layered architecture and counts with social and environmental simulation layers. Social simulation realizes decision making, negotiation, and collaboration. Environmental simulation is dedicated to the diffusion of environmental changes and to the agents' behavior. The system has been applied to fire fighting, for example, where two types of agents of various levels of responsibility (the FireBoss and FireFighter agents) are implied.

To summarize all the papers reviewed, it is possible to classify the following types of environmental information systems (EIS) into three categories:

1. *Local systems.* This type of EIS is a kind of "island solution" which is dedicated to the evaluation or assessment of a few parameters or indicators. In other words, these systems are designed to solve a specific problem. For example, one could find a system that provides a specific assessment of parameters for a specific case study or for a limited area. Domain ontologies for such systems are limited, although they may suffer from possible heterogeneity. As a rule, such systems are effective when working within the application domain but are sensitive to any unforeseen changes.
2. *Multi-functional systems.* These systems provide multiple analysis of input information, can be based upon hybrid techniques, and possess tools and methods of data pre- and post-processing, modeling and simulation. Multi-functional systems are less sensitive to changes in the application domain as they possess tools to manage uncertainty and heterogeneity.
3. *Methodologies/frameworks of EIS development.* Frameworks support all the stages of EIS life cycle, starting with the initial system planning. They include system analysis and domain (problem) analysis phases, and then assist and provide EIS design, coding, testing, implementation, deployment and maintenance. In this case, the consolidated cooperation of specialists from various domains with various backgrounds is necessary. Methodologies/frameworks are based upon interdisciplinary approaches and system analysis.

Table 2.1, Table 2.2, and Table 2.3 show research reports, books and articles that were revised and classified by their functions and goals. Each table is organized in such a way, that the complexity of works cited increases towards the bottom of the table. Many reported works have characteristics of local systems, summed up in Table 2.1 but also propose approaches to more complex and integrated data analysis. The problem lies in the high complexity of domains, which makes the creation of overall general methodologies impossible. These papers are characterized with high flexibility, but knowledge gained can not be easily transferred to other projects. The systems, revised in Table 2.1, are cases of "Flexibility" methodological approach, in agreement with classification introduced by Harmsen [74].

The works presented in Table 2.2 offer more composite approaches, and some of them can exemplify a "controlled flexibility" approach [74], that is a compromise between rigid standardization and ad-hoc engineering. Table 2.3 gathers research reports and articles, which offer methodologies and frameworks for decision making in complex domains. These works give more general view on problem domains, introduce systematization abstractions and methods of control.

To conclude with, it can be said that multiple efforts made for theoretical research and practical implementation of information systems has resulted in many successful applications, some of which, however, lack systemic view and can not offer any coordinated and controlled approach to link methods together into a methodology. In many applications the use of agent-based DSS provides multiple data mining tools and permits discovering systemic properties of domains. One more thing that

Table 2.1 Research papers dedicated to decision support systems in complex systems: Local systems

Ref.	Field of application	Purpose	DM approach
[32]	The author presents results of application of data mining to early intervention for developmentally-delayed children.	To explore the hidden knowledge among medical history data.	Decision trees, association rules.
[102]	The impact of outdoor air pollution on infant mortality.	To confirm that outdoor air pollution contributes to illness and death in adults and children.	Exposure-response functions.
[156]	The authors describe a complex environmental process controlled by an agent-based system.	To improve some previous works that are not based on agents.	Software agents.
[103]	The results obtained from two different neural networks are compared.	To predict traffic related $PM_{2.5}$ and PM_{10} emissions by an artificial neural networks based model.	Feed Forward and Radial Basis Function neural networks.
[165]	The analysis is based on the estimation of the spectral properties of indoor and outdoor pollution particles and the estimation of the integral time scale using the autocorrelation properties of the series.	To investigate the temporal behavior of the indoor and outdoor particles' sizes.	Rotated Principle Component Analysis and Positive Matrix Factorization.
[62]	The definition of a decision support system that deploys an ad hoc agent-based architecture in order to negotiate a distributed diagnostic tool for brain tumors. Implements data mining techniques, transfers clinical data and extracts information.	To improve the quality of brain tumor diagnosis and prognosis.	Multi-agent system.
[145]	In order to evaluate the soil contaminants, a multimedia model is developed and executed for a case-study.	To create soil contamination models and make predictions.	Mathematical modeling.

Table 2.2 Research papers dedicated to decision support systems in complex systems: Multi-functional systems

Ref.	Field of application	Purpose	DM approach
[4]	The paper introduces a multi-agent system for monitoring and assessing air-quality attributes.	The authors offer a solution for continuous surveillance and on-line decision-making.	Software agents.
[79] [76]	The authors present a system for agent-based simulation and support for clinical processes.	To increase the efficiency of hospital process management.	Agent-based simulation.
[124]	It introduces a system capable of belief revision for situation assessment.	To enhance the quality of the situation assessment.	Situation Description Language.
[138] [139]	A distributed approach to the construction of DSS is proposed. The authors present an abstract architecture for a multi-agent decision support system and present case studies.	To facilitate decision making process.	Decision support systems, agent systems.
[174]	The study presents results of research that is dedicated to mining decisions from user profiles in order to find common preferences for different roles of decision makers participating in emergency response operations.	Facilitating a fast and clear description of situation, the generation of effective solutions for situation management, the selection of a right decision maker and supplying him/her necessary data.	Software agents, decision trees.
[22]	This paper reviews the range of approaches to assessment now in use, proposes a framework for integrated environmental health impact assessment, and discusses some of the challenges involved in conducting integrated assessments to support policy.	To bring together existing methods within a more coherent system and to extend these methods in order to provide a more comprehensive methodology for assessing complex, systemic risks and policies.	Risk assessment techniques.
[104]	A decision support system that aims to prepare actors in a crisis situation by a decision-making support system is presented.	To ensure swift and efficient reaction of actors in emergency situations.	Decision support system, software agents.

Table 2.3 Research papers dedicated to decision support systems in complex systems: Methodologies of EIS development

Ref.	Field of application	Purpose	DM approach
[63]	The authors present a methodology for the engineering of agent-based systems, which include the Multi Agent System Development Kit, based on the implementation of the Gaia methodology.	To maintain the integrity of solutions produced at different stages of the development process.	Software agents.
[116]	The author presents an approach to composite decision making for complex domains and demonstrates a number of case studies.	To work out a new approach to complex system study.	Hybrid methods from various disciplines.
[160]	A cross-sectoral approach to assessing sustainable development is introduced.	To create tools able to deal with complex, multi-dimensional phenomena.	Integrated Sustainability Assessment tools.
[191]	An integrated environment for embedding intelligence in newly created agents through the use of Data Mining techniques is presented.	To facilitate dynamic usage of data mining extracted knowledge.	An integrated development framework "Agent" "Academy".
[109]	The author introduces a new engineering discipline "Situational" "Method Engineering" that is aimed in constructing new methods and the associated tools (or in adaption existing ones) to every Information System Development project.	To create new situation-specific methods both by re-engineering of existing method components (so-called "chunks") or by creation of new ones.	Different methods from various disciplines.

should be said about multi-agent systems is that they are designed with agent oriented frameworks, which will be reviewed and discussed in the following section.

2.3   Agent-Based Frameworks for Decision Support Systems

2.3.1   Frameworks for Multi-agent Systems Planning

Agent-oriented software development (AOSD) is a recent contribution to the field of Software Engineering. Agent-oriented methodologies provide a user with necessary theoretical base and a set of practical tools for multi-agent system creation. Besides, they introduce abstractions that support system creation. To date numerous methodologies for agent-oriented software development are being practiced [10], [78], [189], [100], [110] . However, their application to real-world problems is still limited due to their lack of maturity. Evaluating their strengths and weaknesses is an important step towards developing better methodologies in the future. In this connection it seems reasonable to mention that, in spite of a variety of existing methodologies, the most part of them rises from object-oriented methods.

The PASSI methodology has its origins in object-oriented and agent paradigms and unify them using the Unified Modeling Language (UML) notation [50]. It offers a detailed lifecycle development process, that covers stages starting from initial requirements and includes deployment and the social model of agent-based systems [24]. The Tropos methodology puts an emphasis on requirements analysis and, hence, on modeling goals, and on establishing their relations with system entities (actors, tasks, and resources) [58]. It is based on the Belief-Desire-Intention (BDI) agent model [78]. The Prometheus methodology is based on agent-oriented paradigm, and brings rich possibilities for descriptions of agent system in whole and each agent in detail, that can be applied both to BDI and non-BDI agents [192]. MAS-CommonKADS is based on both CommonKADS and object-oriented (OO) methodologies. This enables the developer to build agent-based systems while leveraging the experience of pre-agent methodologies and employing familiar techniques and diagrams [86], [85]. The methodology of fusion inspired creation of the Gaia methodology, which provides an easily-understandable approach [35], [208]. It offers abstractions and diagrams that support multi-agent systems analysis and design phases [208]. The Gaia v.2 is the next version of the Gaia methodology that is more oriented to designing and building systems in complex, open environments [10].

The Rational Unified Process (RUP) process has three "siblings": the methodologies RAP, ADELFE and MESSAGE [108]. ADELFE is oriented to deal with emergent systems and that is why it proposes the usage of cooperative and self-organize agents [11], [12]. The main phases of this methodology are preliminary requirements, final requirements, analysis and design. And, it uses UML and Agent-based Unified Modeling Language (AUML) notations [8], [10]. MESSAGE and INGENIAS assert both RUP-based approach and use the modeling language of Agent-Object-Relationship (AOR) [25], [199], [61], [199], [108], [45], [53]. The first one provides extended models for analysis and design, and the second one propose a meta-model organization of a multi-agent system and provides support on all the stages of system lifecycle. The MAS-CommonKADs methodology heredities from object-oriented methodology Object Modeling Technique (OMT) and extends the CommonKADS models, adapting them to agent-oriented programming. The AAII also takes ideas of the OMT methodology and deepens into relations between the agent and the environment with regards to the interaction model [107]. The AAII inspired the MaSE, which provides environments to support the development stages and can be applied to various heterogeneous domains [38]. Last, the OPEN approach supports design of AgentOPEN agents [195], [77].

The methodologies that were used and studied in depth are discussed in the following sections.

2.3.1.1 The Prometheus Methodology

The Prometheus methodology (see Fig. 2.1) defines a detailed process for specifying, designing and implementing agent-oriented software systems. It consists of three phases [192]:

- "System Specification" phase, which focuses on identifying the goals and basic functionalities of the system, along with the inputs (percepts) and outputs (actions) [162].
- "Architectural Design" phase, which uses the outputs from the previous phase to determine which agent types the system will contain and how they will interact.
- "Detailed Design" phase, which looks at the internals of each agent and how it will accomplish its tasks within the overall system.

System Specification. The Prometheus methodology focuses particularly on identification of goals [110], and description of scenarios [120]. In addition, it requires creation of the functionalities, which are small chunks of behavior related to the identified goals. There is also a focus on how the agent system interfaces with the environment in which it is situated, in terms of percepts that are received from the environment and actions that impact the environment. As part of the interface specification, Prometheus also addresses the interaction with any external data stores or information repositories. The aspects developed in the System Specification phase are: specification of system goals with associated descriptors, development of a set of scenarios that have adequate coverage of the goals, definition of a set of functionalities that are linked to one or more goals and which provide a limited piece of system behavior, and finally, description of the interface between the agent system and the environment in which it is situated.

Achieving goals are central to the functioning of the intelligent software agents. An initial brief system description provides a starting point for building a primary list of goals. The goals are then refined and, after asking "how might this goal be achieved?", the sub-goals of the goal under consideration are given. A group of similar sub-goals provides the basis for each functionality. Functionality is the term used for a "chunk" of behavior, which includes groupings of related goals, percepts, actions, and data relevant to the behavior. Functionalities allow for a mixture of both top-down and bottom-up design. They are identified by a top-down process of goal development, and they provide a bottom-up mechanism for determining the agent types and their responsibilities. Scenarios are complementary to goals in that they show the sequence of steps which take place within the system. Scenarios are used

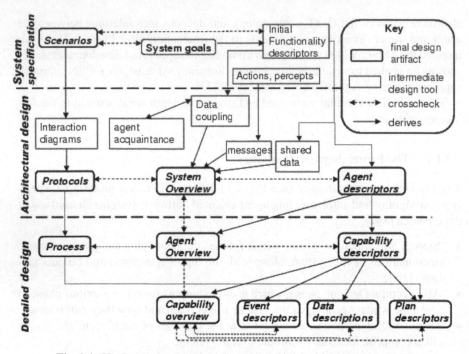

Fig. 2.1 The Prometheus methodology, reproduced with permission [141]

primarily to illustrate the normal operation of the system. As scenarios are developed where there is a need for information from the environment and where actions are required becomes evident. Possible scenario steps include achieving a goal, performing an action, receiving a percept, or referring to another use case scenario.

Agent systems are typically applied in a changing and dynamic environment. The interface description primarily deals with the incoming environment information (percepts) and the mechanisms for affecting the environment (actions). At this stage, a list of percepts and actions is developed. Also, as scenarios and functionalities are developed, it is important to note the data that is produced and used.

Architectural Design. The three aspects that are developed during the Architectural Design phase are: deciding on the agent types used in the application, describing the interactions between agents using interaction diagrams and protocols, and finally describing the system structure through the system overview diagram.

A major decision made during the architectural design is the type of agent used. Agent types are formed by combining functionalities using the criteria of coupling and cohesion. One strong reason for grouping functionalities together is that they use the same data. We do not want agents to know about other agents. Once the agents in the system have been chosen, the pathways of communication (which agent talks to which other agents) as well as the timing of communication (which messages are followed by which other messages) are both identified. The communication is depicted explicitly in interaction diagrams, which are obtained after replacing any functionality with the agent that includes it, and inserting a communication between agents where it is needed. It may that agent be necessary to switch from interaction diagrams to protocols that define exactly which interaction sequences are valid within the system. Finally, the interactions between the agents and the system interface in terms of percepts, actions, and external data are specified. The overall design of the system is thus depicted in the system overview diagram, which brings all the items together.

Detailed Design. In the Detailed Design, for each individual agent, it is decided what capabilities are needed for the agent to fulfil its responsibilities as outlined in the functionalities it contains. The process specifications which indicate more of the internal processing of the individual agents are developed. And when getting into greater detail, the capability descriptions which specify the individual plans, beliefs and events needed within the capabilities are developed. Then, the views that show processing of particular tasks within individual agents are developed. It is during this final phase of detailed design that the methodology becomes specific to agents that use event-triggered plans in order to complete their tasks. As the design develops, what was originally a single functionality may be split into smaller modules, each of which is a capability. For each capability, we need to determine the goals it is to achieve within the system. The agent overview diagram shows the relationships between the capabilities, thus providing a top level view of the agent internals. It also shows the message flow between the capabilities, as well as data internal to the agent. In detailed design, we also want a mechanism which specifies process as well as structure. For this, Prometheus uses a variant of UML activity diagrams, where the activity within a single agent is specified, indicating interaction with other agents via the inclusion of messages within the diagram.

At the final step of the detailed design, each capability is broken down either into further capabilities, or, eventually, into a set of plans which provide the details of how to react to situations or achieve goals. Capability diagrams take a single capability and describe its internals. Prometheus focuses on BDI platforms, which are characterized by a representation that has hierarchical plans with triggers, and a description for each plan that indicates the applicable context. BDI systems choose among the plans that are applicable, and can backtrack in order to try another plan if the one initially chosen is not successful.

At the lowest level of detail, each incoming message to the capability must have one or more plans that respond to that message. Each plan can typically be broken down into a number of sub-tasks, each of which is represented by an internal message. Each plan is triggered by a specific event. This event may be the arrival of a percept, arrival of a message from another agent, or an internal message or sub-task within the agent. If there are several plans that can be triggered by a given event, then it is important to specify the conditions or situations under which the various plans are applicable. This is called context condition. A very important issue associated with the event is the precise specification of the information carried by that event.

2.3.1.2   The Gaia Methodology

The Gaia methodology [208] provides a full support for multi-agent system creation starting from the requirements determination, up to the detailed design. Fig. 2.2 gives a view on organizational diagram of the Gaia methodology [30].

There are two phases of modeling within Gaia: analysis and design. The aim of the first stage is to understand the system structure and its description. The objective of the design stage is "to transform the abstract models derived from the analysis stage into models at a sufficiently low level of abstraction that can be easily implemented".

Analysis phase. It supposes the following steps:

1. Identification of the roles.
2. Detailed description of the roles.
3. Modeling interactions between the roles.

At the first step, as the requirements of the system are stated, two models are created: the Roles model and the Interactions model. To create the Roles model, the developer has to understand the main purposes of the system created. Then, he/she has to analyze the organizational and functional profile of the system, which is decomposed and represented by a set of played roles. The concept of "Role" is one of the key concepts of Gaia methodology, as it determines a function related to some system task (or tasks), which semantically and functionally interacts with the other roles. A role can be related to a system entity. For example, in the case of human organization, a role can represent a "manager" and a "seller". The role is defined by the following attributes: responsibilities, permissions, activities, and protocols.

Responsibilities determine functions of the role and have liveness properties and safety properties. Liveness properties describe the actions and conditions that the agent will bring. In other words, it determines the consequences of the executed procedures, which will be potentially undertaken within a role. Safety properties state crucial environmental conditions, which cannot be exceeded or neglected. Permissions determine the resources and their limits for the role, and are commonly represented by information resources. For example, they can be abilities to read, change, or generate information. Activities are private actions of an agent, which are executed by the agent itself, without communication with the other agents. Every role can be associated with one or more protocols, which state communications with other roles. The described attributes for every role are pooled into a so-called role schemata, thus, comprising the Roles model. The interaction model is focused on protocol description. Protocols determine links between roles and provide the interaction within the multi-agent structure. The protocol definition includes:

- purpose - a brief description or detailed name of the protocol, which discovers the nature of interaction;
- initiator - the name of the role that initiated the interaction;
- responder - the role with which the initiator communicates;

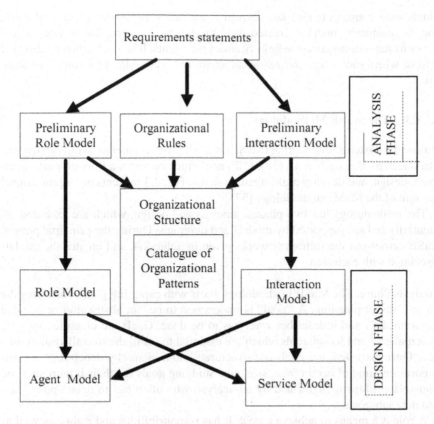

Fig. 2.2 The Gaia methodology, reproduced with permission [30]

- inputs - the information, supplied by the responder during the interaction;
- processing - a brief description of processes realized within the protocol.

Next, the Roles model is created. It contains associated protocols, which comprise the Interaction model.

Design phase. During this phase, Service, Agent and Acquaintance models are created. These models provide a detailed description of the multi-agent system that can then be easily implemented. The Agent model relates roles to every agent type, taking into account that an agent may play one or more roles. The agents form a hierarchy in which the leaf nodes correspond to roles and the other nodes to other agent types. The number of agent instances is also documented. For example, the agent may be called for execution once, or n times, or repeated from m to n times, etc.

The Services model identifies the necessary resources for every function performed by an agent. Every function (or service) has properties, which include inputs, outputs (those are derived from protocols), and pre-conditions and post-conditions, which state constraints and are derived from safety properties of related roles. The Acquaintance model is created from the Interactions and Agent models, and serves to state communication links between the agents. It is represented by directed graphs, where each vertex relates to an agent and every edge to a communication link.

2.3.1.3   The MaSE Methodology

Multiagent Systems Engineering (MaSE) is a full life cycle methodology, which was introduced by DeLoach et al. [38], [10], and which provides a flow of work to analyze, design, and develop multi-agent systems. Fig. 2.3 presents an organizational diagram of the MaSE methodology [37].

The methodology has two phases: analysis and design, which are executed sequentially and are supported by models and diagrams. During the principal phases, MaSE carries out the following works given in Table 2.4, and creates the models associated with each step.

Analysis Phase. The MaSE methodology starts with capturing goals, following the natural way of planning. A "goal" is understood to be "an abstraction of detailed requirements", and it describes *what* has to be done. Goals are classified into the principal goal and its subgoals which are essential to fulfill the overall system mission. Once extracted, the goals are structured into a hierarchy depending on their importance, time of occurrence, size, etc. Studying goals for their importance and inter-relationships is performed by an analyst, who often has to decompose goals into new sub-goals.

A role is a means to achieve a goal. It has responsibilities and rights, as well as functions to describe *how* a goal must be achieved. So, the decomposition offered by MaSE is functional. In MaSE four special types of goals are assigned. They are shown on a Goal Hierarchy Diagram and include summary, partitioned, combined, and non-functional goals. In the "Applying Uses Cases" step, goals are translated into roles and associated tasks. The use cases describe the system behavior, and link roles with events and interactions. These events and interactions are illustrated by Sequence Diagrams that are similar to standard UML sequence diagrams. The Refining Roles step is aimed to specify in more details the links between Goal Hierarchy Diagram and Sequence Diagrams to facilitate their further introduction into MAS. As a rule, the transformation of goals into roles is one-to-one, with each goal mapping to a role, but with possible exceptions. MaSE suggests that roles and goals have to be revised during the Analysis phase. As a result, related roles and goals can often be combined together. Finally, roles are captured in the Role Model, which also shows interactions between role tasks. The Concurrent Task Model contains a schematic representation of the individual tasks for each role in case they are concurrent. Once the concurrent tasks of each role are defined, the Analysis phase is complete.

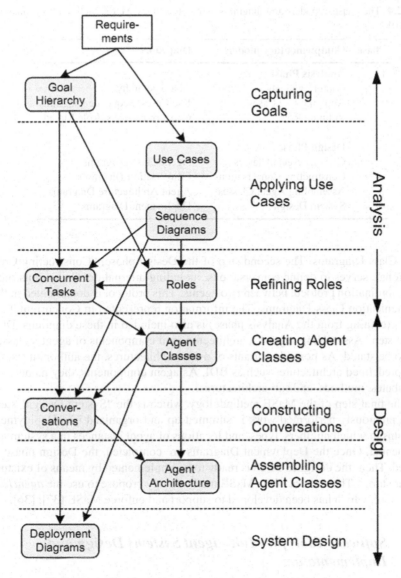

Fig. 2.3 The MaSE methodology, reproduced with permission [37]

Design Phase. The purpose of the "Creating Agent Classes" step is to create agent classes, using the roles defined in the Analysis phase as a basis. The overall organization of agents within the system and their acquaintance are depicted on the Agent Class Diagram. An agent class in MaSE is a kind of a template for a type of agent. It helps in allocating resources and providing communication patterns. During this step each agent is being assigned roles, and communications between agents are settled. The agent classes and conversations are then represented in the

Table 2.4 The phases, models and diagrams associated in the MaSE methodology, adapted from [10]

Phase	Complementary models	Diagrams
1	**Analysis Phase**	
	Capturing Goals	Goal Hierarchy
	Applying Use Cases	Use Cases, Sequence Diagrams
	Refining Roles	Concurrent Tasks, Role Model
2	**Design Phase**	
	Creating Agent Classes	Agent Class Diagrams
	Constructing Conversations	Conversation Diagrams
	Assembling Agent Classes	Agent Architecture Diagrams
	System Design	Deployment Diagrams

Agent Class Diagrams. The second step of the Design phase, "Constructing Conversations", serves to define conversations, including internal details by constructing a coordination protocol between two agents. This protocol is documented in the Communication Class Diagrams. The information represented in Concurrent Task Models (coming from the Analysis phase) is also included in these diagrams. During the step "Assembling Agents", architecture and components of agent´s classes have to be stated. As possible variants of agent architectures, the authors insist on using predefined architectures such as BDI. As agent components, they name a set of attributes, methods, and sub-architectures.

In the final step of the MaSE methodology, which is the "System Design" step, all the previously performed work is summed up and organized into Deployment Diagrams in which numbers, types, and locations of agent instances in a system are documented. Once the Deployment Diagrams are completed, the Design phase is finished. Then, the designed MAS is ready to be implemented by means of existing software tools. The authors of the MaSE methodology propose to use the *agentTool* system tool, which has been developed to support and enforce MaSE [37], [36].

2.3.2   Software Tools for Multi-agent Systems Design and Implementation

Because of the complex nature of problems to solve, multi-agent systems become more complicated to plan and design. New concepts such as goals, roles, plans, interactions, and environment are necessary to identify the system functionality, the interactions between agents, the mental states, and the behavior of the last appeared. On the other hand, MAS have to be secure, mobile and able to manage distributed problem solving. That is why a methodology should facilitate and support agent-based system engineering. This requirement can be achieved if a methodology provides a solid terminology support, precise notations and reliable interactions, and general system functionality organization. Nowadays, there are a number of methodologies for MAS planning and design, which are divided into steps during which the system is firstly described in general terms and then in more details determining the internal functionality of the system entities.

Two well-known methodologies were presented and discussed in the previous section. In this part, software tools for system coding and implementation are discussed. These tools can be viewed as a logical continuation of the methodologies. Indeed, $JACK^{TM}$ Development Environment is related to Prometheus Design Toolkit and MASDK is based on the Gaia methodology.

2.3.2.1 The JACK™Design Tool

The $JACK^{TM}$ Development Environment (referred to as "JACK" from this moment on) is a cross-platform for building, running and integrating multi-agent systems in Java-based environments [91], [189], [158]. The JACK has a visual interface, which supports application creation. This may be done directly in the JACK environment or be imported. For example, it may be imported from Prometheus Development Tool, a graphical editor which provides agent systems design in accordance with its associated methodology Prometheus. JACK enables building applications by providing a visual representation of the system components in two modes: agent mode and team mode [93]. There are several components for the agent-oriented software (AOS) family:

- **JACK ™** is an autonomous systems development platform. It has a proven track record in the development of applications that interact with a complex and ever-changing environment.
- **JACKTeams™**supports the definition of autonomous teams. It supports a wide variety of teaming algorithms, allowing the representation of social relationships and co-ordination between team members.
- **C-BDI™**implements the core of JACK's BDI model in Ada programming language. It is designed for applications where the software needs to be certificated, for example in onboard aviation systems.
- **CoJACK™**is a cognitive architecture used for modeling the variation in human behavior. It is used in simulation systems to underpin virtual actors.
- **Surveillance Agent™**is a JACK-based product that assists surveillance and intelligence agencies in the analysis of behavior patterns.

JACK, written in Java, provides object-oriented programming for the system, encapsulating the desired behavior in modular units so that agents can operate independently. JACK intelligent agents are based on the Believe-Desire Intention model, where autonomous software components (the agents) pursue their given goals (desires), adopting the appropriate plans (intentions) according to their current set of data (beliefs) about the state of the world. Hence, a JACK agent is a software component that has:

- a set of beliefs about the world (its data set),
- a set of events that it will respond to,

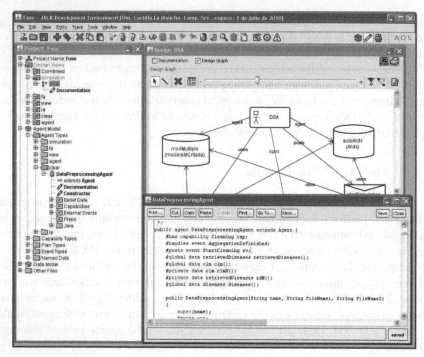

Fig. 2.4 An example of JACK window with Navigator, Diagram design and Programming windows

- a set of goals that it may desire to achieve (either at the request of an external agent, as a consequence of an event, or when one or more of its beliefs change), and,
- a set of plans that describe how it can handle the goals or events that may arise.

JACK permits creation of multiple autonomous agents, which can operate in agent and in team mode within a multi-agent system. MAS creation can be achieved using a graphical interface, as shown in Fig. 2.4. Each JACK project file includes Design views and an Agent and Data model of the application. The Agent model incorporates containers to determine and to code the agents and their capabilities, plans, event types, and belief structures. The JACK^{TM} Agent Language is a super-set of Java. It contains the Java components and syntax while extending it with constructs to represent agent-oriented features. The JACK Agent Compiler pre-processes JACK Agent Language source files and converts them into pure Java, which can then be compiled into Java virtual machine code. The JACK Agent Kernel is the runtime engine for programs written in the JACK Agent Language, which provides a set of agent-oriented classes. JACK extension to team mode enables team models to be treated as peers, and introduces new concepts as team, role, team-data and teamplan. Designing multi-agent systems in this mode requires widened semantics of elements, introducing team reasoning entities,and finally introducing knowledge and internal coordination of the agents within the team.

The key concept which appears here is the role concept. A role defines the means of interacting between a containing team (a role tenderer) and a contained team (a role performer or role filler). In JACK Team mode each team has its lifetime, which is divided into two phases. The first phase is for setting up an initial role obligation structure and the second phase constitutes the actual operation of the team. In addition to the agent beliefs, in team mode, knowledge can be "propagated" over the team members.

The Prometheus Development Kit permits the creation of the skeleton code for its later implementation in JACK, which facilitates the stages of MAS planning and coding. In fact, JACK teams can be used for complex distributed system modeling and problem solving. The JACK Agent Language introduces the main class-level structures which help to create the multi-agent system architecture [92], [93], [94]. These structures are:

1. Agent
2. Capability
3. BeliefSet
4. View
5. Event
6. Plan

JACK Agent Language structures [92] can be categorized as follows:

- Classes (types), which determine functional constructors or units.
- Declarations (#-declarations), which are agent properties. For example, they can identify the usage of plans, the rights to data usage, and define capabilities and queries.
- Reasoning Method Statements (@-statements) describe specific logical guards or logical actions that the agent should perform within the reasoning method. Some of the @-Statements are: @post(parameters), @send (parameters), @parallel (parameters), @wait_for(parameters).

JACK facilitates the creation of flexible platform-independent agent-based software programs. It shares principles of object-oriented programming and offers a usable environment for agent implementation, evaluation and testing.

2.3.2.2 INGENIAS

INGENIAS is an AOSE methodology that embodies theoretical and methodological outlines established in MESSAGE/UML [10]. It is known to extend the basic UML concepts of "class" and "association" with knowledge level agent centric concepts [25], [60], [147]. The INGENIAS Development Kit (IDK) shares a conceptual hierarchy of common elements in MAS specifications and their relations. It widely uses graphical tools that facilitate system design [78]. INGENIAS contains several tools that can be grouped as follows:

1. A visual language for MAS definition, which permits the creation of extensible versions of meta-models of agent systems with a meta-modeling language.

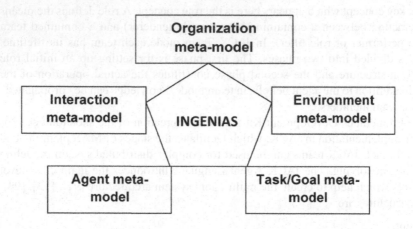

Fig. 2.5 The INGENIAS meta-model, reproduced with permission [164]

2. As one of the ambitions of INGENIAS is to cover all the stages of the industrial
   development process, it offers sets of activities oriented to integrate phases of the
   Unified Software Development Process [95].
3. As a development tool, INGENIAS uses additional tools: a meta-case environment METAEDIT+ that allows building modeling tools and fitting generators
   to application domains and Graph-Object-Property-Role-Relationship (GOPRR)
   meta-modeling language [193], [106].

Fig. 2.5 shows the meta-models that are used in INGENIAS. The focus of the
INGENIAS methodology is to carry out MAS analysis and perform a high-level
design, which is composed as a fusion of coherent meta-models:

1. Agent model, which describes agents and their essential characteristics (tasks,
   goals, states, etc.),
2. Interaction model, which is aimed to describe the interactions between agents.
3. Tasks and goals model, which serves to link together goals and tasks structures,
   allocate information flows and determine mental states for agents.
4. Organization model, which gives a detailed view of the system as a whole,
   putting together elements of the previous two models.
5. Environment model, which describes a domain of interest and the initial information resources.

2.3.2.3 Java Agent DEvelopment Framework

The Java Agent DEvelopment framework (JADE) is a software framework fully implemented in Java language [9]. Developers position JADE as "a middleware for the
development and run-time execution of peer-to-peer applications which are based

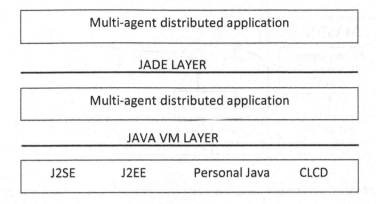

Fig. 2.6 Organization structure of JADE, reproduced with permission [9]

on the agent paradigm and which can seamlessly work and interoperate both in a wired and wireless environment". JADE facilitates the development of distributed applications composed of autonomous entities that need to communicate and collaborate in order to achieve the working of the entire system. JADE is free software and is distributed in open source software under the terms of the Lesser General Public License (LGPL) version 2. Since May 2003, a JADE Board has been created that supervises the management of the JADE Project.

JADE is a run-time system for the Foundation for Intelligent Physical Agents (FIPA) compliant MAS supporting application agents that agree with FIPA specification [48]. Simultaneously, JADE provides object-oriented programming through messaging, agent life-cycle managing, etc. Functionally, JADE provides the basic services necessary for distributed peer-to-peer applications in fixed and mobile environments. Each agent can dynamically discover other agents and is able to communicate with them directly. Each agent is identified by a unique name and provides a set of services, manages them, can control its life cycle and communicates with other agents (see Fig. 2.6).

JADE is a distributed platform that consists of one or more agent containers and is supported by Java Virtual Machine (JVM). JVM provides a complete run time environment for agent execution and allows several agents to simultaneously execute on the same host. The configuration can be controlled via a remote graphic-user interface. The configuration can even be changed at run-time by moving agents from one machine to another, as and when required. JADE has two types of messaging: inter-platform and intra-platform (interacting agents are inside the same platform). Messaging, realized in Agent Communication Language (ACL), is presented in the form of a queue, which can be accessed via a combination of several modes: blocking, polling, timeout and pattern matching based. For a transport mechanism, different protocols can be used: Java RMI, event-notification, HTTP, and IIOP.

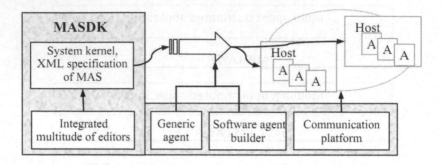

Fig. 2.7 MASDK software tool components and their interaction, reproduced with permission [64]

2.3.2.4   Multi-agent System Development Kit

Multi-agent System Development Kit (MASDK) is a relatively new methodology, created in the Laboratory of Intelligent Systems, St. Petersburg Institute for Informatics and Automation of the Russian Academy of Sciences [64], [65]. The software tool provides support for the whole life cycle of the MAS development. As a terminological foundation the authors use the Gaia methodology [66]. The MASDK 4.0 software tool consists of the following components (also shown in Fig. 2.7):

1. System kernel, which is a data structure for XML-based representation of applied MAS formal specification.
2. Integrated set of the user friendly editors supporting the user's activity aiming at formal specification of an applied MAS under development during the analysis, design and implementation stages.
3. Library of C++ classes of reusable agent components constituting what is usually called the "Generic" agent.
4. Communication platform to be installed in certain computers of a network.
5. "Builder" of software agent instances responsible for generation of C++ source code and executable code of software agents as well as deployment of software agents over already installed communication platforms.

MASDK includes three editors, which act on each of the three levels. The editors of the first level correspond to the Gaia's analysis phase and are dedicated to ontology determination, roles extraction, and determination of protocols and interactions between the agents. The editors of the second level support the design activities and primarily aim for the specification of agent classes. They include agents that determine behavior, agent ontologies, functions and plans. The editors of the third level support the implementation stage of applied MAS and particular components, and lists the agent instances of all classes with references to their locations (hosts names) and initial states of agent beliefs. The next stage corresponds to the design phase of the Gaia methodology, where the developer fills generalized MAS structural entities with internal components, which are the following ones:

1. Fixed (reusable) Generic agent.
2. Meta-model of the agent class's behavior.
3. Multitude of functions of the agent class represented in terms of state machines.
4. Library of additional functions.

The applied MAS specification produced by the designers exploiting the above editors is stored as an XML file in the system kernel. In this specification, including a set of particular components and functions implemented in C++, the Generic agent reusable component forms the input of the software agent builder automatically generating software code based on XSL Transformations (XSLT) technology that permits transforming XML documents into other XML documents.

2.3.3 Comparison of Agent-Based Frameworks

The comparison of agent-oriented methodologies is aimed to thoroughly examine their strengths and weaknesses and determine selecting the most appropriate methodology that can be used for complex systems study. The review includes Prometheus, Gaia, MaSE and INGENIAS. The similarities and differences between these methodologies with regard to their support for MAS lifecycle phases are discussed in [10], [78], [189], [17]. As the methodologies operate with different concepts and notions, the selected features and aspects of their structure and functionality are evaluated and compared. Among the possible features and properties of agent-oriented methodologies which have been compared in several reviews [78], [10], [17], those that could estimate systemic characteristics of MAS were selected for this comparison. They include the following:

1. Lifecycle phase coverage shows what phases of the lifecycle are covered by the methodology.
2. Primitive entities shows the low-level elementary entities the methodology deals with.
3. Agent type indicates whether the methodology supports heterogeneous or homogeneous agents.
4. Team work support shows whether the methodology enables design of agent teams.
5. Complex domains applicability shows whether the methodology is suitable for a complex application domain.
6. Software support shows what software toolkits support the methodology.

The phases are numerated, and:

- "1" stands for "Domain and System Requirements Analysis",
- "2" stands for "Design",
- "3" stands for "Implementation",

Table 2.5 Comparison of agent-oriented methodologies

	Agent-oriented methodology					
	Prometheus	Gaia	MaSE	INGENIAS	ADELFE	TROPOS
Lifecycle phases coverage	1 and 2	1 and 2	1 and 2	1 to 5	1 and 2	1 to 5
Development strategy	Bottom-up	Top-down	Top-down	Hybrid	Top-down	Top-down
Agent type	BDI	Hetero-geneous	Hetero-geneous	Agents with goals and states	BDI	Hetero-geneous
Complex domains applicability	Various, except mobile applications, GPRS, etc.	Wide range of applications	Wide range	Wide range	Emergent systems	Wide range

Table 2.6 Comparison of agent implementation toolkits

	Agent development environment		
	JACK	JADE	MASDK
Environment	Java	Java	C++
Primitive entities	plans	behavior	plans
Agent type	Heterogeneous	BDI	Heterogeneous
Team works support	Yes	Yes	Yes
Application domains	Various heterogeneous domains	Execution state mobility: WWW, mobile application, heterogeneous networks	Heterogeneous environments, including dynamic ubiquitous ones

- "4" stands for "Verification",
- "5" stands for "Maintenance".

Table 2.5 shows the answers received for the methodologies. For the agent implementation environment, JACK, JADE and MASDK were all evaluated. The results of their evaluation is shown together in Table 2.6.

All the revised methodologies were developed for general purposes. For example, the MaSE is intended for closed heterogeneous multi-agent systems development (the number and type of all agents are previously known) and is based on object-oriented approach. The Prometheus is thought to be founded on an agent-oriented paradigm, and it is also a general purpose methodology. INGENIAS has demonstrated that it is the most universal, as it covers all the stages of the lifecycle process. The strong point of the Gaia methodology is that it is easily understood by both specialists and non-specialists. With regard to software support, every methodology utilizes at least one supporting tool and "add-ins". For example, JADE uses a wide number of plug-ins that permit it to integrate with various other applications.

With respect to the complex system study, it seems that each of the reviewed methodologies can be applied, as each of them have been applied to complex domains. However, it seems reasonable that the Prometheus methodology, which is based on the agent-oriented paradigm, is used for this purpose. Another reason for its usage is that it permits the creation of a skeleton code of the multi-agent system for JACK.

Chapter 3
Design and Implementation of the DeciMaS Framework

Make no little plans: they have no magic to stir men´s
blood...make big plans, aim high in hope and work.
Daniel H. Burnham

Abstract. The current chapter offers a complete description of the proposed DeciMaS framework. The chapter shows the components of the DeciMaS framework
and the way in which they are organized. The three DeciMaS framework's stages,
namely, preliminary domain and system analysis, system design and coding, and,
simulation and decision making, are explained in detail. During the first stage, the
analyst researches the following questions: what the system has to do and how it
has to do it. As a result of this collaboration the meta-ontology and the knowledge base appear. The active element of the second stage is a developer, who implements the agent-based system and prepares it for further usage. During the last
phase, the final user, a decision maker, can interact with the system. This interaction consists of constructing solutions and policies, and estimating consequences of
possible actions on the basis of simulation models. Design and implementation of
the system are discussed. The chapter demonstrates how information is transformed
into knowledge and illustrates the transformations made throughout each stage of
the DeciMaS framework. Furthermore, the information about data mining methods,
which are used in the DeciMaS framework, are introduced and depicted in detail.

3.1 Introduction

In this chapter the framework for DSS creation called "Agent Based Framework
for **Deci**sion **Ma**king in Complex **S**ystems" (DeciMaS), which supports the vital
stages of information system creation, is introduced. The structure and the work
flows for each of the stages are defined. Initially, the general view of the DeciMaS
framework is presented, and its phases are briefly explained. Then, an approach for
meta-ontology creation that covers the domain of the problem area as well as the domains related to the DSS itself is described. The mapping procedure for ontologies
will also be demonstrated. Next, each private ontology is examined and its functionality is described as well as its internal and external interactions. Finally, methods
and tools which can be applied at each stage of the MAS execution are suggested.

M.V. Sokolova, A. Fernández-Caballero: Decision Making in Complex Systems, ISRL 30, pp. 47–88.
springerlink.com   © Springer-Verlag Berlin Heidelberg 2012

3.2   The DeciMaS Framework

The purpose of the DeciMaS framework is to provide and to facilitate complex systems analysis, simulation, and their comprehension and management. From this standpoint, and taking into account the results and insights provided in Section 1.3, the principles of the system approach are implemented in this framework. The overall approach used in the DeciMaS framework is straightforward. The system is decomposed into subsystems, and intelligent agents are used to study them. Then the obtained fragments of knowledge are pooled together and general patterns of the system behavioral tendencies are produced [182], [183].

The framework consists of the following three principal phases:

1. **Preliminary domain and system analysis.** This is the initial and preparatory phase where an analyst, in collaboration with experts, studies the domain of interest, extracts entities and discovers its properties and relations. Then, he/she states the main and supplemental goals of the research, and the possible scenarios and functions of the system. During this exploration analysis, the analyst researches the following questions: *what* the system has to do and *how* it has to do it. As a result of this collaboration the meta-ontology and the knowledge base appear. This phase is supported by the Protégé Knowledge Editor, which implements the meta-ontology, and the Prometheus Design Kit, which is used to design the multi-agent system.
2. **System design and coding.** The active "element" of this phase is a developer, who implements the agent-based system and prepares it for further usage. As support at this phase, the JACK Intelligent Agents and JACK Development Environment software tools are used. Once the coding has finished and the system has been tested, the second phase of the DeciMaS is concluded.
3. **Simulation and decision making.** This is the last phase of the DeciMaS framework and it has a very special mission. During this phase, the final user, a decision maker, can interact with the system. This interaction consists of constructing solutions and policies, and estimating consequences of possible actions on the basis of simulation models.

The overall view on the support of the principal phases of the development process within the DeciMaS framework is provided in Fig. 3.1.

3.3   Approach towards Ontology Creation

The main step in organizing the terminological foundation for further analysis and usage (MAS creation, simulation, alarm awareness, etc.) assumes the description of a meta-ontology framework. This step is basic and states the initial quality of further research processes and proper treatment of the concepts. According to Guarino and collaborators [70], an ontology can be understood as an intentional semantic structure encoding the implicit rules constraining the structure of a piece of reality. There are a number of approaches to ontology creation, mostly formed by the specificity

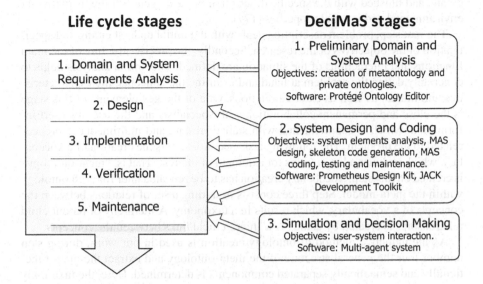

Fig. 3.1 The DeciMaS framework stages

of the domain of interest and the nature of the tasks to be solved (e.g. [163], [14], [197], [23], [68], [6]). The following activities make up an algorithm for ontology creation:

1. Situation description in natural language.
2. Controlled vocabulary creation.
3. Taxonomy creation.
4. Distributed meta-ontology structure creation.
5. Domain of interest ontology statement.
6. Description of tasks to solve and creation of the respective private ontologies.
7. Description of MAS roles and agents, and creation of the system architecture ontology.
8. Description of the agent ontology.
9. Agent environment ontology statement by specifying interaction and communication protocols.
10. Ontologies mapping.
11. Data base filling for a MAS.
12. Data sources delivering to agents.

The given algorithm extends from initial system requirements analysis and situation description towards complete system organization. It is a kind of "bottom-up" inductive method, where global aims are first identified, and then described in more details and finished with the specific description of each agent, stating its particular environmental and internal properties [177].

The first step, problem description deals with the initial understanding and specification of the objectives of the research. Secondly, the precise structure of the main functions and interactions of the situation are defined. This initial analysis helps to concretely define the problem at hand and examine the concepts, their characteristics and mutual as well as external relations. One of the key elements of this stage is the close and permanent collaboration with specialists and the use of expert information, which is supplemented with statistical data and multimedia references related to the problem. The following task (second step) is the creation of a vocabulary, which includes a list of terms for a domain of interest. That is, "meta-ontology" is created on this step. Vocabulary creation has to be repeated later for each ontology within the meta-model. Step three consists of adding a set of relations between the concepts of a vocabulary, which results in a taxonomy. A hierarchy of parent-child relationships is created by using "is-a" hierarchical links between the concepts.

As the inductive method of ontology creation is used in our work, during step number four the general structure of the meta-ontology and extract the main functionally and semantically separated components is determined. Here, the taxonomy created in step three is expressed in an ontology representation language. The language grammar contains elements and constraints that define the presentation and usage of concepts and representation of knowledge, using formal language semantics to embody rules and complex relations between concepts (statements, assertions, queries, and so on). At steps five to eight, private ontologies are created for the components extracted from the meta-ontology. These are the domains of interest, the MAS architecture, the tasks, the agents, and the interactions. For each of these ontologies we have to comply with tasks (two), (three) and (four), or, in other words, private ontologies are created. At steps nine and ten, the private ontologies are mapped together. Mapping occurs as a result of the existing relationships between concepts, their properties and the ontological semantics.

Then, we fill data bases for a MAS (step eleven). In this stage the model of the meta-ontology, which is composed of the models of the private ontologies is obtained. Ontologies, filled with data, are converted into knowledge bases. Finally, the real data is delivered to agents (step twelve) and the multi-agent-based DSS can start functioning.

3.4   The General Structure of the System

The general structure of the proposed system, which is created in accordance with the DeciMaS framework, is presented on Fig. 3.2. The system is virtually and logically organized into a three-leveled architecture, where each level supports the correspondent phase of the DeciMaS [180].

The system belongs to an organizational type. This means that all the agents within it are responsible for particular tasks. That is why there is for a need of any controlling agent for an external entity. The first layer is dedicated to data retrieval, fusion and pre-processing, the second one discovers knowledge from the data, and

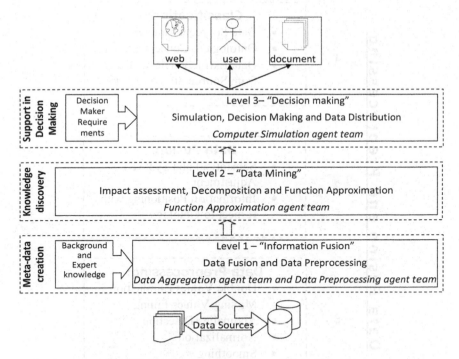

Fig. 3.2 The general view of the system architecture

the third layer deals with making decisions and generating the output information. Let us observe in more details the tasks solved at each level. First, the following work flow takes place:

Information Search - Data Source Classification - Data Fusion - Data Pre-processing - Creation of Beliefs.

Fig. 3.3 shows which data mining methods are executed at the first logical layer. They are grouped into "Classification", "Data Fusion" and "Data Preprocessing" groups. The first and the second groups have to resolve tasks related to information retrieval from data storages and handle the information fusion with respect to domain ontology. The "Data Preprocessing" group includes data clearing tasks and methods.

The second logical level is completely based on autonomous agents, which decide how to analyze data using their abilities to do so. The principal tasks to be solved at this stage are:

- Examining interrelations between the factors and factor groups, revealing hidden patterns and making conclusions about further directions of data analysis.

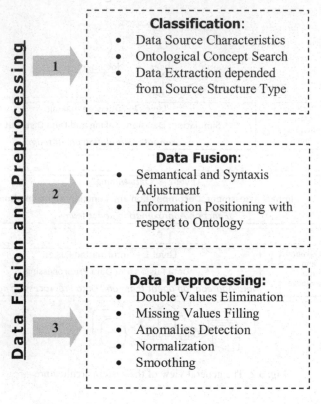

Fig. 3.3 The workflow of the first logical layer

- Creating the models that explain dependencies between factors and their groups, evaluating models and selecting the best ones.

Thus, the aim of the second logical level is to discover the knowledge in the form of models, dependencies and associations from the pre-processed information, which comes from the previous logical layer. The work flow at this level includes the following tasks:

State Input and Output Information Flows - Create Models - Assess Impact - Evaluate Models - Select Models - Display the Results

Data mining methods which are executed at this level are shown in Fig. 3.4. They include methods for classification and prediction.

The third level of the system is dedicated to decision generation. Both the decision making mechanisms and the human-computer interactions are important here. The system works in a cooperative manner, and it allows decision makers to modify, refine or complete the decision suggestions, providing them to the system and validating them. This process of decision improvement is repeated indefinitely until the consolidated solution is generated. The work flow is represented as follows:

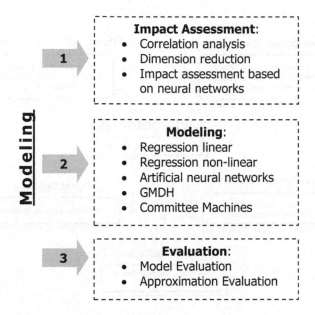

Fig. 3.4 The workflow of the second logical layer

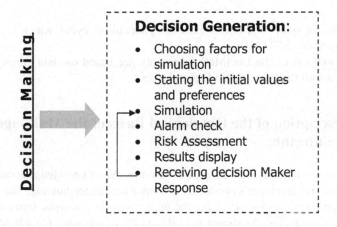

Fig. 3.5 The workflow of the third logical layer

State Factors for Simulation - State the Values of Factors - Simulate - Evaluate Results - Check Possible Risk - Display the Results - Receive Decision Maker Response - Simulate - Evaluate Results - Check Possible Risk - Display the Results

Fig. 3.5 depicts methods for decision making. It illustrates the nature of the human-computer interaction during the decision elaboration process. It is repeated while a user changes initial values and tries various alternatives. This process is

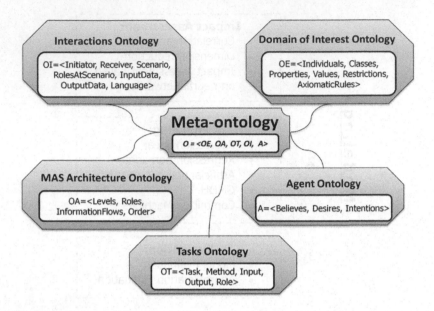

Fig. 3.6 The components of the distributed meta-ontology

indicated by an arrow that forms "decision generation" cycle, which is also called "sensitivity analysis" [126].

The workflows of the DeciMaS framework are based on data mining methods that carry out all the tasks within logical layers.

3.5   Description of the Ontological Basis of the Multi-agent Architecture

It is well-known that ontology creation is based on expert knowledge about the problem area and the developer's previous experience and understanding. The more generalized approach implies extracting the main group of concepts, semantically and functionally connected. As shown in [146], a typical ontology for a MAS includes the following models: domain of interest, aims and tasks, agents, interaction, and environment. Having accepted this model of distributed meta-ontology creation, the structure shown in Fig. 3.6 is proposed as a framework for meta-ontological MAS design.

This meta-ontology model specifies the private ontologies and gives opportunities to generalize about the MAS and the problem area. In the following subsections the focus is set on the components of the proposed meta-ontology.

3.5.1   The Domain of Interest Ontology

When defining the ontology, O, in terms of an algebraic system, we have the following three attributes:

$$O = (C, R, \Omega) \tag{3.1}$$

where C is a set of concepts, R is a set of relations between the concepts, and Ω is a set of rules. Formula (3.1) proposes that the ontology for the domain of interest (or the problem ontology) may be described by offering proper meanings to C, R and Ω.

As we have used the ontology editing software Protégé [151], [161], [136] after widening the formula for our system, we get the following specialization (see formula (3.2)) for equation (3.1) :

$$OE = < Individuals, Classes, Properties, Values, Restrictions, AxiomaticRules >$$
$$\tag{3.2}$$

where

- *Individuals* represent unit exemplars (any given entities), which form the same class. Individuals from the same class share properties and restrictions put on them.
- *Classes* are interpreted as "sets containing individuals", and are organized in a taxonomy in accordance with the hierarchical superclass-subclass relations.
- *Properties* are binary relations on individuals, which enable asserting facts about classes and individuals and can be functional, inverse functional, symmetric, or transitive. The properties are used in restrictions and in axioms.
- *Values* contain the values that can be assigned to individuals.
- *Restrictions* state the permitted and extreme ranges. Generally speaking, they impose constraints on the properties of the classes.
- *AxiomaticRules* use restrictions, boolean algebra and some other concepts such as general classes to create properties and class axioms.

3.5.2   The MAS Architecture Ontology

The initial analysis of the system is carried out with the Gaia methodology [208] and results in revealing and describing the system roles and protocols. The ontology for MAS architecture is stated as:

$$OA = < Levels, Roles, InformationFlows, Order > \tag{3.3}$$

where

- *Levels* correspond to logical levels of the MAS (see Fig. 3.2),
- *Roles* is a set of determined roles,

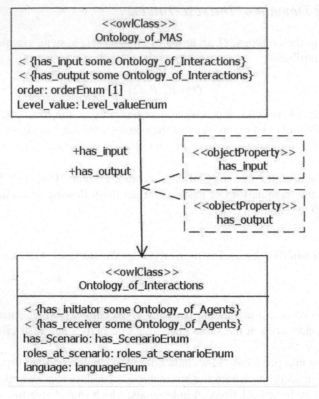

Fig. 3.7 The MAS Ontology

- *InformationFlows* is a set of the corresponding input and output information, represented by protocols,
- *Order* determine the sequence of execution for every role.

The system consists of three levels. The first level is targeted for meta-data creation, the second one is responsible for hidden knowledge discovering, and the third level provides real-time decision support making, data distribution and visualization. This architecture satisfies all the required criteria for decision support systems as it includes the necessary procedures and functions.

In Fig. 3.7 there is an ontology created in accordance with the given formal description (see equation 3.3), which also presents its connection with the Interaction ontology (see Section 3.5.5) determining the protocols. The first level is named "Information Fusion" and it acquires data from diverse sources and preprocesses the initial information to be ready for further analysis. The second layer is called "Data Mining", and there are three roles at this level dedicated to knowledge discovery through modeling. The third level, "Decision Making", carries out a set of procedures including model evaluation, computer simulation, decision making and forecasting based on the models created on the previous level (see Fig. 3.2). The main function of this level is to provide a user - actually, a person who makes decisions - with the possibility to run online real-time "what - if" scenarios.

The end-user, the person making decisions, interacts with the MAS through a System-User Interaction protocol, which is responsible for human-computer interaction. The user chooses the indicator he wants to examine and initiates a computer simulation.

3.5.3 The Tasks Ontology

In order to accomplish the assigned goals, the MAS has to achieve a set of tasks and subtasks. The task ontology is represented by the following components:

$$OT = < Task, Method, Input, Output, Role > \qquad (3.4)$$

where:

- *Task* is a set of tasks to be solved in the MAS,
- *Method* is a set of activities related to the concrete task,
- *Input* and *Output* are input and output data flows,
- *Role* is a set of roles.

In Fig. 3.8 there is a private ontology created in Protégé, which demonstrates the formal model of the Task Ontology. The component "Role" is inherited from the Ontology of MAS Architecture. The tasks are shared and can be accomplished independently, in accordance with an order, which is inherited from the MAS Architecture Ontology through the *Role* component. The task delegation is being delivered for every type of agent. In fact, all the types of agents solve particular tasks and have determined responsibilities. Our system may be considered an organizational MAS [200], which is strictly organized and does not require any kind of control agents.

3.5.4 The Agent Ontology

In our approach, agents are modeled as BDI agents. Their architecture consists of three data structures: *Beliefs*, *Desires* and *Intentions* (which include a plan library). The *Beliefs* are usually represented as facts or in the form of information files and databases, and correspond to the information that the agent has about its environment. *Desires* are actions or goals that the agent wants to achieve, and *Intentions* are the desires that the agent chooses under the given circumstances. *Intentions* are realized in the form of actions, which are formed in a plan library consisting of sequences of steps that the agent can execute to achieve its goals. *Intentions* is a

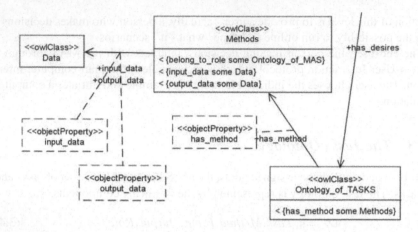

Fig. 3.8 The Tasks Ontology

subset of *Desires*. Hence, we describe every agent as a composition of the following components:

$$Agent = < Beliefs, Desires, Intentions > \qquad (3.5)$$

Each agent has a detailed description in accordance with the given ontology, which is offered in the form of BDI cards, in which the pre-conditions and post-conditions of agent execution, explaining the necessary conditions and resources for the agent's successful execution, are stated. There is also a collaborator, in case there are two or more agents needed to solve the task.

The Agent Ontology created is represented in Fig. 3.9 with general details that include its components. The *Beliefs* contain entities from the *Data* ontology class, which determines the information data resources needed for every agent. The class *Desires* include methods, stored in the Task Ontology, and the class *Intentions* contains entities from the class *Desires* for every activity or task specification.

3.5.5   The Interactions Ontology

The interactions between agents include an initiator and a receiver, a scenario and the roles taken by the interacting agents,the input and output information and a common communication language. The private ontology is setup as:

$$OI = < Initiator, Receiver, Scenario, RolesAtScenario, InputData, OutputData, Language >$$
$$(3.6)$$

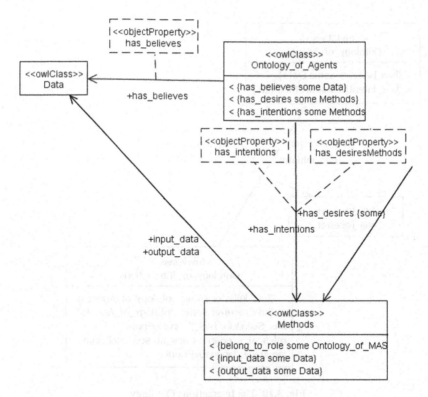

Fig. 3.9 The Agent Ontology

where

- *Initiator* and *Receiver* are roles, which are assigned to split the information and deliver it to the proper agents,
- *Scenario* corresponds to a protocol,
- *RolesAtScenario* is a set of roles that the agents play during the interaction,
- *InputData* and *OutputData* are represented by informational resources, read and created, respectively,
- *Language* determines the communication language.

The graphical representation of the protocol (depicted in Fig. 3.10) shows the main components, their properties and connections with other sub-ontologies.

3.5.6 The Distributed Meta-Ontology

The Distributed Meta-Ontology is obtained as a result of private ontologies mapping, and is pooled by their common use and execution. Existing relations between concepts, their properties and the ontological semantics make mapping possible.

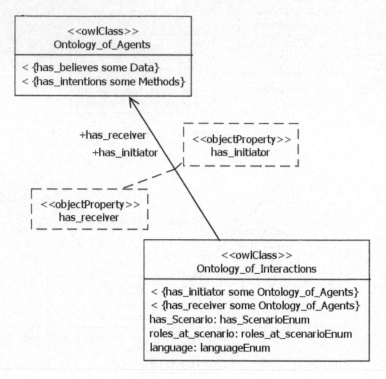

Fig. 3.10 The Interactions Ontology

The shared ontological dimension, filled with the data, provides agents with correct addressing of proper concepts and synchronizes the MAS functionality. The problems that appear at this stage are mostly associated with data heterogeneity. Indeed, in many cases data might be stored in different sources, represented by various identifiers and measured unequally. These procedures can be solved by different methods.

3.6   Data Mining Methods in the DeciMaS Framework

Large amounts of raw data make up a complex system, but not all the information can be of practical use, because a lot of information does not contain valuable patterns and knowledge about the system. That is why there is a necessity to use data mining methods, which help to extract knowledge from factual information, presented in data sets. Information is transformed from the initial "raw" state to the "knowledge" state, which suggests organized data sets, models and dependencies, and, finally, to the "new knowledge" state, which contains recommendations, risk assessment values and forecasts. The way the information changes as it is passes through the DeciMaS stages, is shown in Fig. 3.11.

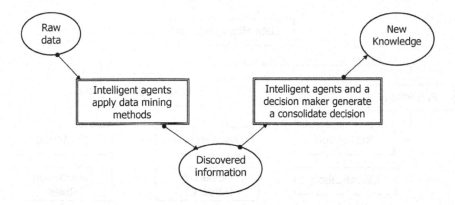

Fig. 3.11 Information transformation from the "raw" to the "new knowledge" state

The rest of the chapter is dedicated to the description of data mining methods, which are used in the DeciMaS framework and which provide its functionality. In general terms, data mining refers to extracting or "mining" knowledge from data sources. However, data mining can be viewed as a part of the knowledge discovery process which consists of an iterative sequence of the following steps [73]:

1. Data cleaning (to remove noise and inconsistent data).
2. Data integration (where multiple data sources may be combined).
3. Data selection (where data relevant to the analysis task are retrieved from the database).
4. Data transformation (where data are transformed or consolidated into forms appropriate for mining by performing summary or aggregation operations, for instance).
5. Data mining (an essential process where intelligent methods are applied in order to extract data patterns).
6. Pattern evaluation (to identify the truly interesting patterns representing knowledge based on some interestingness measures).
7. Knowledge presentation (where visualization and knowledge representation techniques are used to present the retrieved knowledge to the user).

Actually, steps from 1 to 4 are a consequence of data preprocessing procedures, which are performed to prepare data for mining itself. In accordance with [148], data mining methods can be classified and presented in form of a hierarchy depending on the kind of pattern (see Fig. 3.12).

In agreement with the classification hierarchy provided on Fig. 3.12, data mining methods can be divided into tools for *prediction* and *knowledge discovery*. Various methods originated in statistics, probabilistic theory, fuzzy logics, evolution analysis, and methods of artificial intelligence can be used for classification and prediction. The *knowledge discovery* has more objectives and that is why it uses methods

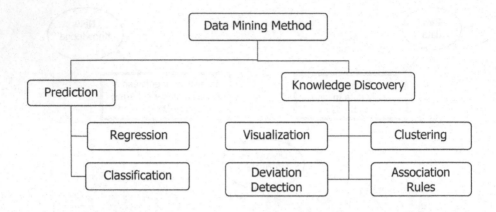

Fig. 3.12 Classification of data mining methods, adapted from [148]

and tools from statistics, cluster analysis, decision rules, computer-user interaction and others.

The DeciMaS framework uses various data mining techniques to support knowledge transformation during each phase:

- The first phase is information preprocessing. This is a standard phase of every knowledge discovery process. As a rule, primary data are incomplete, noisy and inconsistent. Attributes of interest may not be available and various errors (of data transition, measurement, etc.) may occur. A considerable number of data mining algorithms are susceptible to the quality of the data. That is why data mining techniques usually improve accuracy and efficiency of mining algorithms [73], [128], [34]. We discuss here data clearing methods, missing values and outliers detection and correction, methods for normalization, and integration or fusion [18].
- The second phase is knowledge discovery. Here the data mining methods used are revised. The outline for this section is as follows. We begin with bivariate and multivariate statistical approaches for estimation and prediction [112], [132]. These methods include linear simple and multiple regression and its non-linear analogues. Then, the methods used for CS decomposition and partitioning [111] are presented. Next, methods of artificial intelligence, such as artificial neural networks (ANN) are used [28]. The ANN are trained with different training algorithms such as backpropagation (BP), resilient propagation (RPROP) and training with genetic algorithms (GA) [75], [173]. Then, self-organization methods, in particular, Group Method of Data Handling (GMDH) for function approximation and prediction [43] are discussed. Finally, committee machines (CM), which enable obtaining complex hybrid models and forming composite decisions [75], are presented.

- The third phase deals with knowledge presentation and decision making. For these objectives, in the DeciMaS framework, methods for decision theory are used. We describe the techniques used for alternative generation and ranking, and sensitivity analysis. With respect to user-computer interaction, visualization methods, which are allowed by programming environments, are used.

3.6.1 Methods for Information Fusion and Preprocessing

3.6.1.1 Preprocessing Initial Data

Having briefly described the main analogies and differences between data mining and exploratory analysis, the exploratory data analysis methods are described in detail in this section. We begin by focusing on univariate exploratory analysis. This phase is often fundamental to understand what might be discovered during a data mining analysis. It often reveals problems with data quality, such as missing items and irregular data.

3.6.1.2 Identification of Outliers

Data fusing and further cleaning compose the preparation phase for data mining [73], [19]. The consistency of obtained data series is checked, and, first of all, outliers are identified. "Outliers are extreme values that lie near the limits of the data range or go against the trend of the remaining data" [111]. The presence of outliers harms the execution of some data mining methods and, in most cases, outliers are errors of data measurement or extraction.

The best known method of outliers identification is the Z-score standardization [111], which sets a value as an outlier if it is out of the $[-3\sigma, 3\sigma]$ interval of the standard deviation. Since this method is very sensitive to the presence of outliers, we decided to try a more robust statistical method of outlier detection, based on using the interquartile range. This method uses the interquartile range (IQR), which is a measure of variability and is calculated as $IQR = Q3 - Q1$, and represents the spread of the middle 50% of the data. It states that a value is an outlier if:

- it is lower than $(Q1 - 1.5(IQR))$,
- tt is higher than $(Q3 + 1.5(IQR))$.

where $Q1$ is a $25th$ percentile, $Q3$ is a $75th$ percentile.

3.6.1.3   Missing Data Treatment

Missing data is one of the common problems that almost always appears during the initial stage of data preprocessing. The reasons for missing data can be any of the following ones:

- Gaps caused by equipment errors (in case of sensors and other sources of real-time information).
- Gaps caused by the absence of available information (for example, by lacks of factual information in database).
- Gaps that appear as the result of previous data clearing procedures (for example, outlines or double values elimination).

The problem of missing values needs to be resolved, as it skews further data analysis. We need a number of methods and techniques known to be "robust" for missing data [13], [96]. In case of short data sets the fields with missing values cannot be omitted, because each value in the sample is valuable and necessary. As studies say, the absence of data (and information), even after using special techniques, is less favorable than working with complete data sources [111]. In the scientific literature dedicated to this theme, there are several ways to deal with missing data offered [111], [13], [73]. The most common ways are as followed:

1. Replacing the gap with some constant specified by the specialist.
2. Replacing the gap with the field mean.
3. Replacing the gap with a value generated at random from the variable distribution observed.
4. Replacing the gap with the mean of the k neighboring values.
5. Replacing the gap with the golden ratio of the neighboring values.
6. Using Bayesian methods to calculate the value to replace the gap.
7. Replacing the gap with the value received as a simulation result.
8. Using a multiple imputation technique.

When looking through the methods listed above, the advantages and disadvantages of each of them could be analyzed:

- "Replacing the gap with some constant" is one of the simplest methods, which requires previous expert advice. Moreover, it can fail in case of dynamic and changeable data series.
- The second method, which uses the field mean, is not optimal as the mean may not always be the best choice for what constitutes a "typical" value. For example, if many missing values are replaced with the mean, the resulting confidence levels for statistical inference will be overoptimistic, since measures of spread will be reduced artificially [111]. It must be stressed that replacing missing values is a gamble, and the benefits must be weighed against the possible invalidity of the results [73].
- "Replacing with the mean of the k neighboring values", on the contrary, is more adequate, because this method is more local and the means for various gaps would differ.

- The next method of the missing values treatment, "replacing a gap with a value generated at random from the variable distribution observed", leads to more variability in the data sample, but, the tendency of the data series would not be kept, since randomly chosen values would tend to make it weaker.
- "The golden ratio" as the value to fill the gap is an alternative to the mean. The golden ratio is 0.62 percent of the value of the previous and 0.38 percent of the next to the gap value. Its application has the same advantages and disadvantages as when using the mean.
- Some more sophisticated and complex methods (6, 7 and 8 in the enumeration list) of missing data treatment are explained in detail in [31], [167].
- In agreement with one of the most powerful techniques (the sixth in the enumeration list), which includes Bayesian methods, a unified framework is provided for addressing all the problems of survey inference, including missing data [31].
- Another approach deals with the "simulation of posterior distributions". For example, simulation by Markov chain is frequently used, as it "is a natural companion and complement to the current tools for handling missing data, and, in particular, the expectation-maximization (EM) algorithm" [167].
- Moreover, simulation facilitates "inference by multiple imputation". "Multiple imputation is a technique in which each missing value is replaced by $m > 1$ simulated values [119]. The m sets of imputations reflect the uncertainty about the true values of the missing data. After the multiple imputations are created, m plausible versions of the complete data exist. Each of them are analyzed by standard data methods. The results of the m analysis are then combined to produce a single inferential statement (e.g. a confidence interval or a p-value) that includes uncertainty due to missing data" [119].

3.6.1.4 Smoothing

The next task for data preprocessing is smoothing. The aim of smoothing is to reduce irregularities and fluctuations in data. Smoothing of data is one of the basic data preprocessing procedures, and it is particularly recommended when the data contains errors or is being corrected and filled with new values after missing and erroneous data treatment. The basic and simplest technique is moving-average smoothing. Its more advanced modifications, such as centered and weighted moving averages as well as exponential smoothing, are also widely used in practice [143].

- *Centered moving average* is a modification of a simple moving average in which the average is placed in the middle of an interval of n periods, i.e. at the $n/2$ point, which holds for odd numbers n.
- *Weighted moving average* is an averaging algorithm that limits the participation of individual observations according to their "age".
- *Exponential smoothing* is a technique that uses exponentially decaying coefficients for the n previous values.

The simple moving average method uses the following formula to calculate a value x (with the number n of previously used observations):

$$x(t+1) = \frac{x(t) + x(t-1) + \ldots + x(t-n)}{n} \tag{3.7}$$

In the case of weighted moving average, the formula as shown in reference (3.7) changes to formula (3.8), as the weight to each past value is added:

$$x_w(t+1) = w_1 \times x(t) + w_2 \times x(t-1) + \ldots + w_n \times x(t-n) \tag{3.8}$$

Exponential smoothing can be viewed as a kind of weighted moving average that produces a smoothed statistic when two observations are available. It is particularly recommended for short-time forecasting [143]. The raw data sequence is represented by x_t and the output of the exponential smoothing algorithm is written as s_t. The following algorithm is used:

$$s_t = \alpha x_t + (1 - \alpha) \times s_{t-1} \tag{3.9}$$

where s_t is the exponentially smoothed value, x_t is the observed value at the same point of time, α is the smoothing constant value, and s_{t-1} is the previous exponentially smoothed value. The steps to this process are as follows:

1. Set smoothed values as s_t.
2. Set value to α, where $0 < \alpha < 1$.
3. For the first step initialize $s_0 = x_0$.
4. Calculate each smoothed value as shown in formula 3.9, where $t = 0, 1, \ldots, n$, being n the number of values in a set.

The values of α which are close to one have a lower smoothing effect and give a greater weight to recent changes in the data, while values of α that are closer to zero have a greater smoothing effect and are less responsive to recent changes. Reference [143] states that the value α can be calculated directly from the previous data values and it also can differ with each iteration in order to obtain better results. For this reason, the authors suggest using the sum squared prediction errors for different values of α.

3.6.1.5   Normalization

In this work, various data mining techniques to process and transform the initial information have been used. Many of them have particular requirements regarding the variables' ranges. As a rule, ranges of variables differ greatly, and are not comparable with each other. That is why they cannot be calculated and analyzed together. On the other hand, if studying two variables and one has greater values than the other, often times, the first variable will have more influence on the final answer. That is why this step of data preprocessing procedure is dedicated to reviewing the normalization techniques, which are used in our work. In agreement with [111] and

[75], two of the more prevalent methods were used. We take X as the original field value and X^* as the normalized field value.

Z-score standardization works by taking the difference between the field value and the field mean value and scaling this difference by the standard deviation of the field values. It can be calculated as:

$$X^* = \frac{X - mean(X)}{SD(X)} \qquad (3.10)$$

where $mean(X)$ is the mean and $SD(X)$ is the standard deviation.

The $Min - Max$ normalization evaluates whether the field value is greater than the minimum value $min(X)$ and then measures this difference by the range. The $Min - Max$ normalization is calculated as:

$$X^* = \frac{X - min(X)}{max(X) - min(X)} \qquad (3.11)$$

where $min(X)$ and $max(X)$ refer to the minimum and maximum values, respectively, of the variable field. The $Min - Max$ normalization values will range from zero to one, which is very useful for certain data mining techniques, for example when working with artificial neural networks as cited in reference [75].

3.6.1.6 Correlation Analysis

The presence of collinearity implies that there is some missing information, as one or more of the collinear factors is redundant and adds no new information. Correlation between variables can interfere and disrupt the correct execution of data mining procedures and lead to false results. That is why correlation analysis has been included as one of the stages of data preprocessing. The mutual correlation between the variables of a model, for both dependent and independent variables, provoke unstable and unreliable results. It occurs because the variables interact with each other and impede the calculation of adequate model coefficients - for example, regression, artificial neural networks, and so on. In DeciMaS parametric and non-parametric correlation for various data series and data mining techniques are used.

Parametric correlation is used for long data samples. We calculate linear correlation using the equation for the *Pearson correlation coefficient*:

$$r(X,Y) = \frac{Cov(X,Y)}{\sigma(X)\sigma(Y)} \qquad (3.12)$$

where $Cov(X,Y)$ is the coefficient of covariation between X and Y, and $\sigma(X)$ and $\sigma(Y)$ indicate the standard deviations of variables X and Y, respectively. The covariance $Cov(X,Y)$ is a measure of the concordance between the variables of interest. It can be calculated as:

$$Cov(X,Y) = \frac{1}{N} \sum_{i=1}^{N} [x_i - \mu(X)] [y_i - \mu(Y)] \tag{3.13}$$

where $\mu(X)$ and $\mu(Y)$ indicate the mean of the variables X and Y, respectively.

Non-parametric correlation. The general prerequisites for the evaluation of non-parametric correlation are both noisy and uncleared data and short datasets. When working with real applications, we can face situations where data sources are limited and short. This is why some non-parametric rank correlation coefficients have been included in a number of data mining methods used in the DeciMaS framework. The *Kendall non-parametric correlation coefficient* τ is defined as:

$$\tau = \frac{n_c - n_d}{0.5n(n-1)} \tag{3.14}$$

where n_c is the number of concordant pairs, and n_d is the number of discordant pairs in the dataset. A concordant pair is a pair of a two-variable (bivariate) observation datasets X_1, Y_1 and X_2, Y_2, where:

$$sgn(X_2 - X_1) = sgn(Y_2 - Y_1)$$

The *Spearman's rank correlation coefficient* is the other statistic which is calculated for non-parametric correlation. To calculate the Spearman´s coefficient datasets Xi and Yi have to be converted into rankings xi and yi and sorted. Then, the differences d_i between the ranks of each observation are calculated. The Spearman´s coefficient is written as:

$$\rho = 1 - \frac{6\sum d_i^2}{n \times (n^2 - 1)} \tag{3.15}$$

where $d_i = x_i - y_i$ is the difference between the ranks of corresponding values X_i and Y_i, and n is the number of values in each data set. If there are no tied ranks in either Xi or Yi, then the Spearman correlation coefficient formula (3.15) and the Pearson correlation coefficient formula (3.12) obtain identical results.

3.6.2 Methods for Knowledge Discovery

In the first stage, data is retrieved in accordance with an ontology and the data understanding and data preparation phases are completed. Exploratory data analysis has been completed and descriptive information has been gathered. Having completed the data mining methods of the "Information Retrieval and Fusion" phase, homogeneous data free from outliers and double values can be collected.

3.6.2.1 Statistical Approaches for Estimation and Prediction

One of the mostly frequent and essential tasks in knowledge discovery is function approximation. Regression is one of the most powerful and simple techniques for modeling. The most common types of regression are shown in Fig. 3.13.

In practice, the choice of the fitting curve can be made graphically, analytically, or via experimentation. The first method is acceptable for solo cases, but in cases of multiple and numerous computer-based studies the second and third methods have preference. For the DeciMaS framework the following types of regression models are calculated:

a) $y = a + bx$;
b) $y = a + b/x$;
c) $y = a + x^b$;
d) $y = a + bx + cx^2$;
e) $y = a + bx_i + cx^2 + dx^3$;
f) $y = a + b^x$;

Univariate regression. In linear simple regression, dependent variable (or response) Y and independent (or explanatory) variable X is used, and Y is described as a function of X. The linear regression model is based on paired observations (x_i, y_i) and can be expressed by a regression function:

$$y_i = a + bx_i + e_i, i = 1, 2, \ldots, n, \tag{3.16}$$

where a is the intercept of the regression function, b is the slope coefficient of the regression function, and e_i is the random error of the regression function, relative to the ith observation. The first two components of the regression function as shown in formula (3.16), which are a and bx_i, which describe the linear function that fits the paired observations:

$$\widehat{y_i} = a + b \times x_i, i = 1, 2, \ldots, n, \tag{3.17}$$

where $\widehat{y_i}$ denotes the fitted ith value of the dependent variable, calculated on the basis of the ith value of the explanatory variable, x_i. The error term e_i describes how well the regression line approximates the observed response variable. Fig. 3.14 gives view to a graphical representation of the simple linear regression model.

In order to obtain the analytic expression for the regression line, it is sufficient to calculate the parameters a and b on the basis of the available data. The method of least squares is often used for this purpose:

$$b = \frac{\sum x_i y_i / n - \sum y_i \sum x_i / n^2}{\sum x_i^2 / n - (\sum x_i / n)^2} \tag{3.18}$$

$$a = \sum \frac{y_i}{n} - b \sum \frac{x_i}{n} \tag{3.19}$$

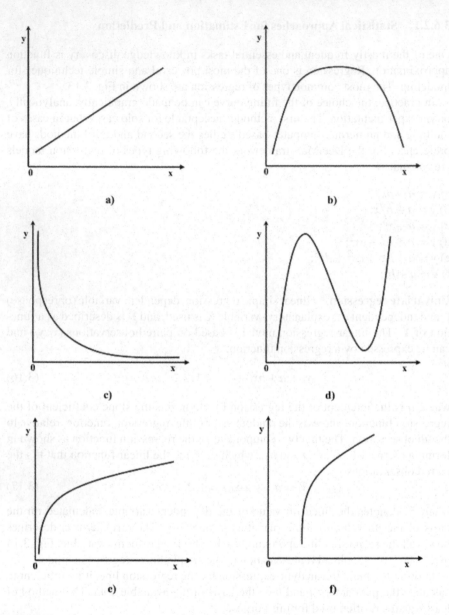

Fig. 3.13 Most commonly used types of regression lines for function approximation, adapted from [41]

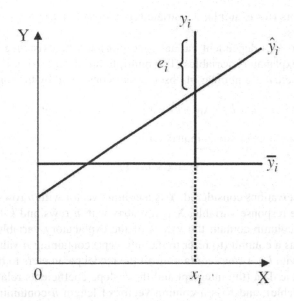

Fig. 3.14 Representation of the regression line, adapted from [57]

The most common equations for univariate regression can vary with the form of the model:

$$y = a + bx \qquad (3.20)$$
$$y = a + b/x \qquad (3.21)$$
$$y = a + b\sqrt{x} \qquad (3.22)$$
$$y = e^{a+bx} \qquad (3.23)$$

Although the first equation describes linear regression, the next ones are typical examples of non-linear regression. To obtain coefficients a and b for the regression of types b) and c), additional variables are used. For example, in the case of b) regression, the variable $z = 1/x$ is introduced and the formula $y = a + b/x$; transforms into its linear analogue $y = a + bz$, which can be solved with formulas referenced as equation (3.18) and equation (3.19). The process of solving the c) case is similar, if an additional variable $z = \sqrt{x}$ is introduced. The regression model d) is transformed into linear form $\ln(y) = a + bx$ and is then solved.

After having been created, the regression models are evaluated and can then be used for further calculations.

Multiple regression. In real case studies, situations arise in which there is more than one explanatory variable, and therefore the method of multiple regression must be selected. Multiple regression is used when it is impossible to neglect all of the influencing factors (for example, univariate regression) to adequately describe dependent variables.

Let us state Y as a dependent variable and consider the existence of a matrix that contains k explanatory variables. The multiple linear regression is defined for $i = 1, 2, ..., n$, where n is a number of observations considered by the formula:

$$y_i = a + b_1 \times x_{i1} + b_2 \times x_{i2} + ... + b_k \times x_{ik} + e_i \qquad (3.24)$$

or, equivalently, in more compact matrix terms,

$$\mathbf{Y} = \mathbf{X} \times b + \mathbf{E}, \qquad (3.25)$$

where, for n observations considered, $\mathbf{Y}$ is a column vector with n rows containing the values of the response variable, $\mathbf{X}$ is a matrix with n rows and $k + 1$ columns which for each column contain the values of the explanatory variables for the n observations, plus a column (to refer to the intercept) containing n values equal to 1; b is a vector with $k + 1$ rows containing all the model parameters to be estimated on the basis of the data (the intercept and the k slope coefficients relative to each explanatory variable); and $\mathbf{E}$ is a column vector of length n containing the error terms [57]. Such a plane is defined by the following equation:

$$\widehat{y}_i = a + b_1 \times x_{i1} + b_2 \times x_{i2} + ... + b_k \times x_{ik} \qquad (3.26)$$

In order to determine the fitted plane, it is necessary to estimate the vector of the parameters $(a, b1, ..., bk)$ on the basis of the available data. We can obtain a solution in a similar way as bivariate regression. In terms of matrices it is given by $\mathbf{Y} = \mathbf{X}\beta$, where:

$$\beta = (\mathbf{X}'X)^{-1}\mathbf{X}'Y \qquad (3.27)$$

The explanatory variables, which are included into the model, must be independent from each other. In other words, they must not correlate with each other. In accordance with some recommendation [41], correlation between the variables may not exceed 0.7.

Another difficulty whith including explanatory variables in a multiple regression model is the existence of multi-collinearity. To calculate multi-collinearity we use the determinant of the matrix, which consists of the *Pearson correlation coefficients*. Let us consider having three independent explanatory variables. Then, the determinant *Det* of the matrix will be the following:

$$Det\,|R| = \begin{vmatrix} r_{x_1x_1} & r_{x_2x_1} & r_{x_3x_1} \\ r_{x_1x_2} & r_{x_2x_2} & r_{x_3x_2} \\ r_{x_1x_3} & r_{x_2x_3} & r_{x_3x_3} \end{vmatrix} = \begin{vmatrix} 1 & 0 & 0 \\ 0 & 1 & 0 \\ 0 & 0 & 1 \end{vmatrix} = 1,$$

as all the correlation coefficients $r_{x_ix_j}$, where $i = 1, 2, 3$ and $j = 1, 2, 3$, are equal to zero. In the opposite case, the multi-collinearity between explanatory variables can be detected if the determinant equals zero:

$$Det\,|R| = \begin{vmatrix} 1\ 0\ 0 \\ 0\ 1\ 0 \\ 0\ 0\ 1 \end{vmatrix} = 0.$$

A value $[n-1-1/6(2m+5)\lg(Det\,|R|)]$ is accepted as a criterion for multi-collinearity. If it exceeds the standard value χ^2 with $1/2n(n-1)$, then the explanatory variables cannot be included in the same model.

3.6.2.2 Dimension Reduction Based on Sensitivity Analysis

An alternative approach to the multidimensional visualization of data is the reduction of the principal components [57]. The main goal is to reveal the relationships among influential factors by studying each explicative input variable on dependent output variables of the ANN. It has been demonstrated that the neural network-based sensitivity analysis strategy not only reveals the relative contribution of each explicative variable on dependent variables but also indicates the predictability of each dependent variable [187], [26].

Sensitivity analysis provides a measure of the relative importance among the inputs of the neural model and illustrates how the model´s outputs vary in response to the variation of an input. The sensitivity analysis is executed for a previously trained network model of feed-forward type. The network has several inputs, which are set to explanatory independent variables X, and the output is set to dependent variable Y. The method consists of the following sequential steps:

1. The output of the network is calculated.
2. The chosen input is set to zero while the other inputs are given values of independent variables X.
3. The output of the network is calculated with the absence of one input.
4. The received value of the output is compared to the output from step 1.
5. The difference received at Step 4 is recorded.

This process is repeated for each input, and all dependent variables of the network are ranked in descending order of the obtained value. The variables with the highest values are supposed to be the most influencing inputs.

3.6.2.3 Group Method of Data Handling

Group Method of Data Handling (GMDH) was developed in 1968, and is an inductive self-organizing method, which is applied to complex domains for data mining and knowledge discovery purposes (e.g. [89], [52], [166], [135]). The method is aimed to obtain a mathematical model of the object or of the process. It is generally used for identification and process forecasting [125], [19], [90]. The general optimization formula solved by the GMDH algorithm is the following:

$$\tilde{g} = arg \underbrace{min}_{g \subset G} CR(g), \; CR = f(P, S, z^2, T, V) \qquad (3.28)$$

where G is the set of considered models; CR is an external criterion of model g quality from this set; P is the number of variables in the set; S is the model complexity; z^2 is the noise dispersion; T is the number of data sample transformations; V is the type of reference function. We are working under the assumptions that z^2, T, and V are constant, and the formula given above can be rewritten as one-dimensional:

$$CR(g) = f(S) \qquad (3.29)$$

The method is based on a selection procedure. It consists of sequentially repeated steps, where model-candidates are tested and compared with the given criterion. The model has a multi-layer form. Each successive layer increases the order of the lower layer polynomial. The general connection between input and output variables is expressed by the Volterra functional series, a discrete analogue of which is the Kolmogorov-Gabor polynomial:

$$y = a_0 + \sum_{i=1}^{m} a_i \times x_i + \sum_{i=1}^{m}\sum_{j=1}^{m} a_{ij} \times x_i \times x_j + \sum_{i=1}^{m}\sum_{j=1}^{m}\sum_{k=1}^{m} a_{ijk} \times x_i \times x_j \times x_k \ldots \quad (3.30)$$

where $X(x_1, x_2, \ldots, x_m,)$ is the input variables vector and $A(a_1, a_2, \ldots, a_m,)$ is the vector of coefficients or weights. The family of GMDH algorithms counts with different algorithms. The basic GMDH algorithm is the parametric combinatorial algorithm COMBI [125]. First, the algorithm calculates the models of following types:

$$y = a_0 + a_1 \times x_i, \; i = 1, 2, \ldots, M \qquad (3.31)$$

Next, the models are evaluated with the the the cross-validation criterion $PRR(s)$:

$$PRR(s) = 1/N \sum_{i=1}^{N} (y_i - y_i(B))^2, \qquad (3.32)$$

In the second phase, models have a more complicated form:

$$y = a_0 + a_1 \times x_i + a_2 \times x_j, \; i = 1, 2, \ldots, F, \; j = 1, 2, \ldots, M, \; F \leq M \qquad (3.33)$$

where F is the number of models that have the best value of criterion, and are selected for the next round. In the third phase, models get one more additional component:

$$y = a_0 + a_1 \times x_i + a_2 \times x_j + a_3 \times x_k, \; i = 1, 2, \ldots, F, \; j = 1, 2, \ldots, F, \; k = 1, 2, \ldots, M \qquad (3.34)$$

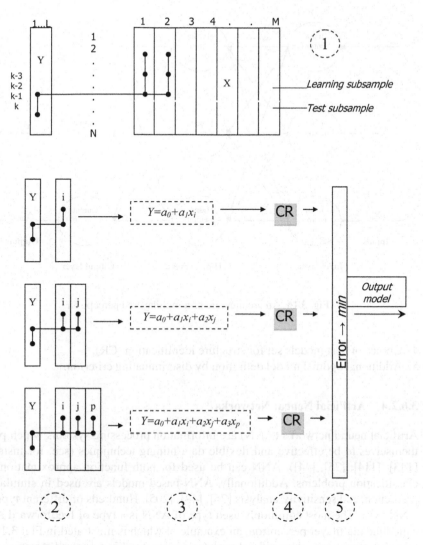

Fig. 3.15 The layered structure of GMDH algorithm, adapted from [69]

and the algorithm goes on this way until the minimum of the criterion is achieved. The models are evaluated by the given criterion, and the procedure is carried on for as long as the minimum of the criterion is found. Fig. 3.15 illustrates a general view of a GMDH model created with the COMBI (combination) algorithm.

The numbers in Fig. 3.15 represent the steps of the COMBI algorithm execution:

1. Data sampling.
2. Creation of partial description layers.
3. Creation of partial descriptions.

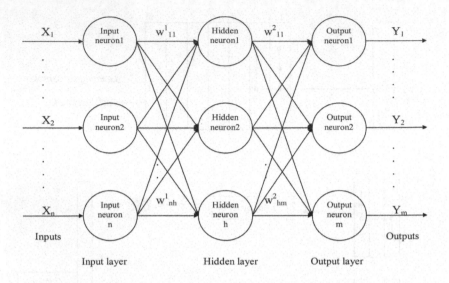

Fig. 3.16 An architecture of a multilayered perceptron

4. Choice of best models set for structure identification (CR).
5. Additional optimal model definition by discriminating criterion.

3.6.2.4 Artificial Neural Networks

Artificial neural networks (ANN) are information processing systems which prove
themselves to be effective and flexible data mining techniques (see, for instance,
[143], [114], [75], [44]). ANN can be used for both function approximation and
classification problems. Additionally, ANN-based models are used in simulation,
prediction, and sensitivity analysis [75], [44], [205]. Hundreds of different types of
ANN exist. The most commonly used type of ANN is a type of feedforward ANN
called the multilayer perceptron, an example of which is illustrated in Fig. 3.16. In
this type of network, the artificial neurons are arranged in a layered structure containing an input layer, a "hidden" layer (or more than one layer) and an output layer.

The backpropagation algorithm approximates the weights of the neural network
[201]. In fact, the backpropagation (BP) algorithm is the basic algorithm for multilayered networks learning. It uses the method of gradient descent and looks for the
minimum of the error function in weight space. During the running of the algorithm, the weights are constantly adjusted until the combination of weights which
minimizes the error function is found. Since this method requires the computation
of the gradient of the error function at each iteration, the activation function has to

be continuous and differentiable. The activation function more frequently used for this purpose is the sigmoid function:

$$F(x) = \frac{1}{(1+e^{-\alpha x})} \tag{3.35}$$

The first derivative of sigmoid function has the following form:

$$F(x) = \alpha F(x)(1 - F(x)) \tag{3.36}$$

The BP algorithm consists of two general phases, namely feedforward and back-propagation [157]. The stopping criterion is commonly defined as:

$$E = \frac{1}{2}\sum_{i=1}^{p}|Y_i - OUT_i|^2 \tag{3.37}$$

where Y is an ideal outcome of the neural network, and OUT is the output that a network gives at the given iteration for a provided training pair. The backpropagation of the errors from the output to the input changes the weights of the connections. This minimizes the error vector. Briefly, the algorithm can be described as:

1. Feed-forward computation. Set W weights to small initial values. For every node calculate the output:

$$NET = \sum_{i=1}^{n} w_{j,i}x_i, \tag{3.38}$$

$$F(node) = \frac{1}{(1+e^{-\alpha NET})}, \tag{3.39}$$

   where n is the number of inputs. Calculate outcomes of the network.
2. Backpropagation of the output layer. Calculate the backpropagation error:

$$\delta_k = out_k(1 - out_k)(y_k - out_k), \tag{3.40}$$

   where k is the number of the output nodes, out_k is the output of the node k and y_k is the desired output for the current training pair.
3. Backpropagation to the hidden layer. Calculate backpropagation errors for each hidden node using the following formula:

$$\delta_j = out_j \times (1 - out_j)\sum(\delta_k w_{j,k}), \tag{3.41}$$

   where k is the node of the posterior and j is the node of previous layer.
4. Update weights using the following formulas:

$$\Delta w_{i,j} = \alpha \Delta w_{i,j} + (1 - \alpha)\eta \delta_j out_i, \tag{3.42}$$

   where i is the previous and j is the posterior layer, η is a learning rate parameter, $0 \le \eta \le 1$, and α is a momentum term $0 \le \alpha \le 1$, that prevents the training algorithm from stopping at a local minima.

$$w_{i,j} = w_{i,j} + \Delta w_{i,j}.$$

The training stops when the stopping criterion is met.

Resilient propagation algorithm (RPROP) is an adaptive technique based on the standard backpropagation algorithm. The backpropagation algorithm modifies the weights of the partial derivatives. RPROP does not count on the value of the partial derivative. It only considers the sign of the derivative to indicate the direction of the weight update. The basic principle of RPROP is to eliminate the harmful influence of the size of the partial derivative in the weight step.

$$\Delta w_{ij} = -sign\left(\partial E / \partial w_{ij}\right) \times \Delta_{ij} \tag{3.43}$$

Δ_{ij} is an update value. The size of the weight change is determined exclusively by this weight-specific "update value". Δ_{ij} evolves during the learning process based on its local sight on the error function E, according to the following learning-rule. The weight update Δw_{ij} follows a simple rule:

• If the derivative is positive (increasing error), the weight is decreased by its "update value".
• If the derivative is negative, the "update value" is added.
• If the partial derivative changes its sign, (i.e. the previous step was too large and the minimum was missed) the previous "weight update" is reverted.

In comparison to all other algorithms, only the sign of the derivative is used to perform learning and adaptation. The size of the derivative decreases exponentially with the distance between the weight and the output layer. When using RPROP, the size of the weight step is dependent only on the sequence of signs, and learning is spread equally over the entire network.

3.6.2.5   Genetic Algorithms

Genetic algorithms (GA) were invented by Holland [80], who describes this idea in his book, "Adaptation in Natural and Artificial Systems". GA were discovered as a useful tool for search and optimization problems, and they are used to search for the best solution (e.g., the minimum for optimization tasks). GA belong to the class of evolutionary computing algorithms and have recently gained more ground thanks to their multiple applications, especially in the case of complex problems (see [170], [118], [82]). Generally, a GA is a non-linear optimization method for global heuristic search that uses ideas of natural process of chromosomes reproduction, mutation and selection [57]. A GA handles a population of possible solutions, where each solution is represented with a chromosome. Fig. 3.17 shows a solution phenotype and a corresponding chromosome. The factors are related to genes in chromosomes. Fig. 3.18 illustrates the "population" term.

A chromosome is an abstract representation of a coded solution that may consist of "0"s and "1"s. The string structures in the chromosomes undergo operations

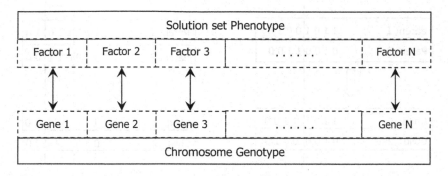

Fig. 3.17 The solution phenotype, adapted from [173]

Population		
Chromosome 1	1 0 0 1 1 1 1 1
Chromosome 2	1 0 0 0 1 0 0 1
Chromosome 3	0 1 0 1 0 0 1 1
Chromosome 4	0 0 0 0 0 1 1 1

Fig. 3.18 Population, adapted from [173]

similar to the natural-evolution process which yield increasingly optimal solutions. The quality of the solution is based on a "fitness" value, which relates to the objective function for the optimization problem. These operations are called genetic operators. They imitate natural processes that occur in populations of living entities. The genetic operators include crossover, mutation, reproduction, and selection.

The crossover operator is applied to the fitting chromosomes, and it is responsible for exchanging information from the "parents". During the crossover, the chromosomes swap their parts of genetic code, creating new individuals. Generally speaking, there are two types of crossover: one-point crossover and n-point crossover. The one-point crossover involves the random selection of a crossover point on the genetic code, while the n-point crossover involves the selection of n points on the genetic code to swap (see Fig. 3.19). The mutation operator is capable of generating new "genetic" information. It acts with a specific probability, and changes a gene (or several genes of a chromosome) to opposite values. The mutation operator can prevent the population from converging and "stagnating" at local optima (see Fig. 3.20). The reproduction operators are applied directly to the chromosomes, and they are used to perform mutation and recombination over the solutions of the problem. The selection is done by using a *fitness function* (see Fig. 3.21. Each chromosome has an associated value corresponding to the fitness of the solution that it represents. After creating a population, a fitness value is calculated for each member in the population, as each chromosome is a candidate solution. The fitness should correspond

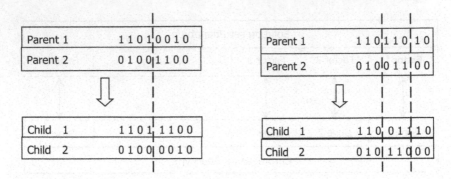

Fig. 3.19 1-point and 2-point crossover, adapted from [173].

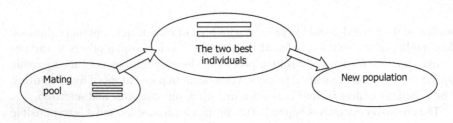

Fig. 3.20 Mutation, adapted from [173].

Fig. 3.21 Selection procedure, adapted from [173].

to an evaluation of how proper the candidate solution is. The optimal solution is the one which maximizes the fitness function [143].

The genetic algorithm process is discussed through the GA cycle provided in Fig. 3.22. Reproduction is the process by which the genetic material in two or more parents is combined to obtain one or more offspring. In the fitness evaluation step, the individual's quality is assessed. Mutation is performed to one individual to produce a new version of it, where some of the original genetic material has been randomly changed. The selection process helps to decide which individuals are to be used for reproduction and mutation in order to produce new search points.

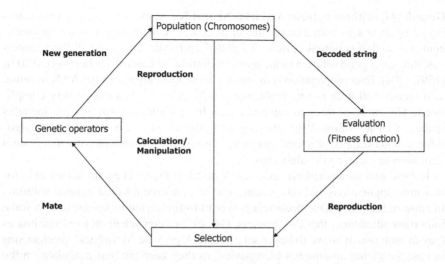

Fig. 3.22 Genetic algorithm cycle, adapted from [173].

In general, a genetic algorithm consists of three mechanisms [173]:

1. Parent selection
2. Genetic operation for the production of descendants (offspring)
3. Replacement of parents by their descendants

In more detail, these mechanisms can be viewed as follows (see also [173], [57]):

1. Create a population of n chromosomes (suitable solutions for the problem).
2. Evaluate the fitness $f(x)$ of each chromosome x in the population.
3. Breed a new generation of chromosomes through the following genetic operations:

   - *Selection*: select two parent chromosomes from a population according to their fitness (the better the fitness, the bigger the chance to get selected).
   - *Crossover*: with a crossover probability, crossover parents to form new offspring (children). If no crossover is performed, the offspring is the exact copy of its parents.
   - *Mutation*: with a mutation probability, mutate new offspring at each locus (position in the chromosome).
   - *Acceptance*: place the new offspring in the new population

4. Evaluate the fitness for each individual of the offspring.
5. Replace the worst ranked part of the population with the offspring.
6. If the end condition is satisfied, stop, and return the best solution in the current population; if not, repeat all the steps starting from step 2.

Genetic Algorithms to train Artificial Neural Networks. GA are good at exploring complex spaces with a large number of components, as they provide an intelligent way to find solutions close to the global optimum. Thanks to all their properties, they are a good solution to the problem of training feedforward networks [203], [159], [204]. Backpropagation is the most widely used algorithm for ANN training, as it works well with simple problems. It fails quickly when data is very complicated. The curve of error function does not have a simple form, since it includes spaces of local minima. When moving along this curve, the backpropagation algorithm can get trapped at a local minima, although the introduction of a momentum term aims to escaping the algorithm.

In turn, genetic algorithms suffer a few disadvantages. They are not suitable for real-time applications and take a long time to converge into the optimal solution. In spite of this, they have become a powerful optimization technique due to some important advantages that they possess. GA are not dependent on local minima as they monotonously move to better solutions. GA provide "drawback" mechanisms in case the global minima has been passed, as they keep the best candidates in the population [209]. When combined with neural networks, a GA is used as a training algorithm, where a matrix of internal weights of the neural network is represented by a population of binary strings [159]. In this case, the genes of the chromosome represent weights, and with every new offspring population the error function is moved down until it approaches the optimal or acceptable solution, by changing the matrix of internal weights.

As we work with short data sets, we do not have to optimize and make the structure of the networks more complex. However, genetic algorithms are used for ANN training. The parameters to initialize network training with GA are, as a rule:

- Population size
- Subpopulation chosen to mate
- Mutation order (one, two-point mutation)
- Stopping criteria (fitness function)

The fitness function should show how well a neural network can approximate a given data set. That is why statistical criterion is chosen, as follows:

$$fitness = 1/2 \frac{\Sigma(y_i^2 - \widetilde{y}_i^2)}{n}, i = 0, 1, \ldots, n. \tag{3.44}$$

where y_i is an ideal value and $\widetilde{y}_i$ is approximated by the neural network value for the $i-th$ step.

3.6.2.6 Committee Machines

A committee machine (CM) is a type of hybrid module that consists of several models. Practically, it represents a "committee" of experts (or models, which play expert roles). The response of a committee is always a composition of each expert's response [26], [75]. A CM provides accurate results, but it is not "visible" and is

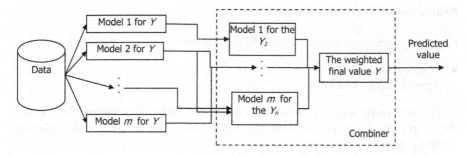

Fig. 3.23 The general architecture of committee machines, adapted from [75]

poorly understood by a human decision maker. The final response of the CM, as it is shown in Fig. 3.23, can be written as:

$$Y = \frac{\sum_{i=1}^{N} Y_i Cr_i^2}{\sum_{i=1}^{N} Cr_i^2} \qquad (3.45)$$

where Y is the response of the CM, Y_i is the output of the i-th model, and Cr_i is the criterion for the given i-th model.

3.6.2.7 Evaluation of Models

Model Comparison

Fitting a model to data can be interpreted as a way of replacing a set of observed data values with a set of estimated values obtained during the fitting process. The number of parameters in the model is generally much lower than the number of observations in the data. We can use these estimated values to predict future values of the response variables from future values of the explanatory variables.

The Cross-Validation

The cross-validation method is one of the most popular techniques for model evaluation. The idea of the method is to divide the sample into two sub-samples. They consist of a training sample having $n - m$ observations and a validation sample having m observations. The first sample is used to fit a model and the second one is used to estimate the expected discrepancy of the model.

With respect to the selection of the number of the observations in the validation data set m, there is a practical recommendation that suppose 75 per cent and 25 per cent for the training and validation data sets, respectively [57], are followed.

Evaluation of Regression Models

Regarding regression models, a two-phase evaluation process is followed:

- Model evaluation
- Approximation evaluation

The first phase means evaluation of the model fitting to a sample data. Here, the main index for proper fit of the regression line is calculated. It is the index of determination R^2, defined as:

$$R^2 = \frac{\sigma_{\hat{Y}}^2}{\sigma_Y^2} = 1 - \frac{\sigma_E^2}{\sigma_Y^2} \tag{3.46}$$

where $\sigma_{\hat{Y}}^2$ represents the explained variance, σ_Y^2 stands for the unexplained variance, and σ_E^2 stands for the residual variance. The coefficient R^2 is equivalent to the square of the linear correlation coefficient, and its value lies within the interval $[0; 1]$. In general, a high value of R^2 indicates that the dependent variable Y can be accurately predicted by a linear function of X [57]. The statistical significance can be checked by an F-test of the overall fit, and then t-tests of individual parameters can be carried out. Thus, to evaluate R^2 the F-test, which calculates the F-statistic, is used:

$$F = \frac{R^2}{1 - R^2} \frac{n - m - 1}{m} \tag{3.47}$$

where n is a number of values in a sample; m is an X-parameter (in the case of linear regression, it is equal to the number of independent variables, X). This F-statistic follows the F-distribution under the null hypothesis. To decide if the hypothesis is accepted or not, we compare the F-statistic with the critical value of F_{tabl} at level α with $df_{between} = m - 1$ and $df_{within} = n - m - 1$ [41] degrees of freedom. If $F > F_{tabl}$, the results are significant at level α.

In the case of multiple regression, a corrected index of determination, R^2, is calculated (see formula 3.46). This is defined as:

$$\bar{R}^2 = 1 - (1 - R^2) \frac{n - 1}{n - m - 1}, \tag{3.48}$$

and then tested with F-statistic (see formula 3.47).

In order to test if regression coefficients a and b are significant in the case of linear regression, t-test statistics are calculated as:

$$t_a = \frac{a}{m_a}, \tag{3.49}$$

$$t_b = \frac{b}{m_b}, \tag{3.50}$$

where

$$m_a = \sqrt{\frac{(y-\hat{y})^2}{n-2}\frac{\sum x^2}{n\sum(x-\bar{x})^2}}, \tag{3.51}$$

$$m_b = \sqrt{\frac{(y-\hat{y})^2/(n-2)}{\sum(x-\bar{x})^2}}. \tag{3.52}$$

The values t_a and t_b have a t-distribution with $n-2$ degrees of freedom if the null hypothesis is true. Comparing them with the critical t_{tabl} with $df = n-2$ degrees of freedom at level α, we can conclude that coefficients a and b are significant with $t > t_{tabl}$.

For the case of multiple regression, coefficients $b_i, i = 0, \ldots m$ (see equation (3.52)) can be evaluated as:

$$m_{b_i} = \frac{\sigma_y\sqrt{1-R^2_{yx_1\ldots x_p}}}{\sigma_{x_i}\sqrt{1-R^2_{yx_{i}x_1\ldots x_p}}}\frac{1}{\sqrt{n-m-1}} \tag{3.53}$$

where σ_y is the mean squared error for Y, σ_{x_i} is the mean squared error for factor x_i, $R^2_{yx_1\ldots x_p}$ is the determination coefficient for the given regression, $R^2_{yx_ix_1\ldots x_p}$ is the determination coefficient of dependence between the factor x_i and all the other factors of the regression model. The t-score is calculated with formula 3.50.

Approximation evaluation deals with estimating some statistics for the initial and for the approximated data sets. It is based on the idea that if the model fits the sample data set, their variances should be equal. For this purpose the Bartlett's test is used [175]. Under the hypothesis, H_0, that all samples have equal variances, the statistics are calculated as:

$$X^2 = \frac{(N-k)\ln(S_p^2) - \sum_{i=1}^k(n_i-1)\ln(S_i^2)}{1+\frac{1}{3(k-1)}\left(\sum_{i=1}^k\left(\frac{1}{n_i-1}\right)-\frac{1}{N-k}\right)}. \tag{3.54}$$

where k is the number of samples, n_i is the size of the ith group, $N = \sum_{i=1}^k n_i$, and $S_p^2 = \frac{1}{N-k}\sum_i(n_i-1)S_i^2$ stand for the pooled variance, which is a weighted average of the group variances. The test statistic has approximately a χ^2_{k-1} distribution. The H_0 hypothesis is rejected if $X^2 > \chi^2_{k-1,\alpha}$, where $\chi^2_{k-1,\alpha}$ is the upper tail critical value for the χ^2_{k-1} distribution.

Evaluation of Artificial Neural Networks

One of the great problems of neural network-based algorithms is that there are no specific statistics to evaluate them. In practice, artificial intelligence borrows evaluation tools from statistics [75]. Apart from the training error and the number of epochs, there is a limited number of statistical criteria applied. The most common indicators are the loss functions:

1. Absolute error:
$$AE = |y_i - \widehat{y}_i|,\qquad(3.55)$$

2. Squared error:
$$SE = (y_i - \widehat{y}_i)^2,\qquad(3.56)$$

3. Mean absolute error:
$$MAE = \sum_{i=1}^{N} \frac{|y_i - \widehat{y}_i|}{N},\qquad(3.57)$$

4. Mean Squared Error:
$$MSE = \sum_{i=1}^{N} \frac{(y_i - \widehat{y}_i)^2}{N},\qquad(3.58)$$

The next group of criteria for ANN evaluation includes correlation and determination coefficients (see formulas 3.12 and 3.46 respectively), calculated for initial and approximated data sets.

3.6.3   Methods for Decision Generation

The family of decision making and supporting systems is very large, and includes many types of DSS, which can be grouped into [172], [168]:

1. Depending on collaboration type:

   - Individual DSS
   - Group support systems
   - Virtual teams

2. Decision making methodology:

   - Decision theory
   - Multi-objective programming

Collaborative support systems are widely distributed computer-based DSS, which are able to maintain support of organizational decisions, starting with individual and finishing with groups. Working personnel compare tasks, and their goals usually overlap. *Group support systems* facilitate collaboration and teamwork, as they overcome time, space and economical limitations and provide informational and knowledge-based resources [40], [176]. As a result of technological impact, *virtual teams* began to appear, which are aimed at solving communication and collaboration problems, for example, in the case of geographically distributed workers. The decision making strategies, that listed, are the most common in scientific research, as it is demonstrated in references [1], [20], [98].

In accordance with Schniederjans [168], decision theory is a field of study that applies mathematical and statistical methodologies to help provide information about which decisions can be made. Decision theory methodologies depend on type of decision environment, which can include:

1. Certainty: the decision maker knows the payoffs of every alternative for every particular state of environment.
2. Risk: information is presented in a probabilistic form or is given partially.
3. Uncertainty: there is no information about future situation.

In practice, there are many approaches and strategies to make choices, and Anderson [1] compares them and introduces the term "ladder of methods". With each step of the ladder, the complexity of decision making criteria becomes more complicated. In the first stage the intuition is used and a list of alternatives based on priority is created.

In the next stage, a set of analytic values is taken into account. Screening is the first of the analytic methods. As we move up the ladder, we pass through *Pluses-and-Minuses decision tables*, which are adequate for most decisions, is a decision table with $+/0/-$ evaluations of the facts entered in the cells. It does not have a numerical evaluation of each alternative, however. For more precise evaluation, on the next stage, we can use a *1-to-10 value table*, where a numerical ranking of each alternative under each state of nature is calculated. The highest step on the *Decision Ladder* is the decision tree/table. This decision table can be used for risky environments when the probabilities of values changes.

The final mission of the system is to draw together possible decisions and recommendations for user. The term "decision" is used here as a process of selecting the best models, and making predictions to comply with the user´s requirements and limitations. The final decision to which choice to make is a prerogative of the specialist. He/she can choose among different alternatives which the system generates. In more detail, he/she can choose between competing models, which differ in their attributes. Before making a final decision, the specialist or a group of specialists involved, have to elaborate a system of criteria in order to be able to evaluate alternative decisions. Then, using the system of criteria, receive criteria for each of the alternatives and rank them. Ranking makes for optimal decision selection, as alternatives can be ranked by desired criterion/criteria.

Table 3.1 Decision planning sheet

Alternative decision a_j	Values $y_i(a_j)$ for the function of interest $Y(a_j)$			
	y_1	y_2	$\ldots$	y_n
a_1	$y_1(a_1)$	$y_1(a_2)$	$\ldots$	$y_1(a_n)$
a_2	$y_2(a_1)$	$y_2(a_2)$	$\ldots$	$y_2(a_n)$
$\ldots$	$\ldots$	$\ldots$	$\ldots$	$\ldots$
a_m	$y_m(a_1)$	$y_m(a_2)$	$\ldots$	$y_m(a_n)$

Table 3.1 shows how alternative decisions can be evaluated by the decision planning sheet. It contains utility values for each alternative. A decision maker may base its choice on analysis and speculation of the sheet.

Next, the user makes a choice, based on the most important criterion. This choice is subjective, and the optimal alternative has the best evaluation for this criterion. The following formula explains how the choice is made:

$$a >\approx b \Longleftrightarrow u(W(a)) \geq u(W(b)), \tag{3.59}$$

where a and b are alternative decisions, W is a value of a criterion, $u(W)$ is utility function, $W(a)$ and $W(b)$ are criterion values for alternatives, $u(W(a))$ and $u(W(b))$ are utility values for values $W(a)$ and $W(b)$ respectively, $\Longleftrightarrow$ is a sign of double implication; $>\approx$ is a sign of loose/unrigorous overbalancing/superiority for alternatives. Formula (3.59) implies that in the case of $a >\approx b$, the utility function for alternative a must have a higher value than that of alternative b.

Dealing with a set of alternatives, A, the best alternative decision, $a*$, is written as:

$$a* : max \, u(W(a)), \tag{3.60}$$

$$a \in A.$$

In our study, we use the outcomes of the variable of interest as the utility criterion, and we convert the outcomes into criterion $W(a)$, with the aim of evaluating how close we are from the optimal decision.

Chapter 4
A Case Study for the DeciMaS Framework

Goals are dreams we convert to plans and take action to fulfill.
Zig Ziglar

Abstract. A case study for environmental impact assessment upon human health is presented in this chapter. Research has been carried out with the aim to demonstrate the possibilities of DeciMaS framework and its ability to manage complex problems. This chapter presents an example of using the DeciMaS with the aim of developing a decision support system for impact assessment upon human health evaluation. Meta-ontology of the domain of interest, the system itself and its mapping are presented in the chapter. The relationship between environment and human health is considered so complex that it has been decided to face it as the case study. Indeed, to date environmental impact assessment is still a hot research topic. In addition, a multi-agent architecture for a decision support system is shown. The meta-ontology of the system, consisting of the models for Ontology of Environment, Ontology of MAS Architecture, Ontology of Agents, Ontology of Interactions and Ontology of Tasks is fully detailed. The sequence of the steps for the DeciMaS framework design with Prometheus Development Kit is introduced. In this sense, the following principal abstractions of the agent-based decision support system are shown: goals and scenarios, actors, roles, data and interaction, and agents. Its implementation with Jack Development Environment is presented as well.

4.1 Introduction

In this chapter, a case study for environmental impact assessment upon human health is presented. We demonstrate how the DeciMaS framework can be applied to a complex and multi-dimensional issue such as an "Environment - Human health" system. The reason this domain was chosen is its high complexity and weak predictability. "Human health" is a composite of many components: physical, chemical, social, psychological, and others. The modeling of its interactions with environmental pollutants is the object to be studied with the DeciMaS framework.

After the brief discussion of the "Environment - Human health" system, the ontology of domain, private ontologies for the components of the meta-ontology and

M.V. Sokolova, A. Fernández-Caballero: Decision Making in Complex Systems, ISRL 30, pp. 89–138.
springerlink.com

their mapping will be shown. In agreement with the meta-ontology, we decompose the functional dimension of the system into logical levels, and discuss their purposes. Then, the MAS design by means of Prometheus methodology is presented. This covers the stages of System and Detailed design, with extraction and definition of scenarios, goals, roles and information flows [182]. Finally, the general system structure and functionality, with the focus on intelligent agents and their implementation in JACK will be presented and discussed.

4.2 Human Health Environmental Impact Assessment

4.2.1 Environment and Human Health

Sustainable development is a broad concept that spans across social, economical, environmental, medical, demographical and political spheres. That is why the abstract and highly complex concept of "sustainable development" includes components (in agreement with the renewed Strategy of Sustainable Development of the European Union [155]), which is related with climate change, sustainable transport, production and consumption, natural resources management, social issues (demography, migration, etc), and public health. Environment is a clear example of a complex domain, composed of numerous self-organized subsystems. If interactions of humans within the environment are studied, the level of complexity of such a system greatly increases [123], [22], [130], [186]. It is a fact that the environment affects human health. Climate changes together with growing anthropogenic impact intensify interactions within the "environment - human health" system.

Humans are affected by this global imbalance, and react with direct and indirect health problems, some examples include "excessive heat-related illnesses, vector- and waterborne diseases, increased exposure to environmental toxins, exacerbation of cardiovascular and respiratory diseases due to declining air quality, and mental health stress. Vulnerability to these health risks will increase as elderly and urban populations increase and are less able to adapt to climate change. In addition, the level of vulnerability to certain health problems vary by location. As a result, strategies to address climate change must include health as a strategic component on a regional level. Improving health while addressing climate change will contribute to public health infrastructure today, while reducing the negative consequences of a changing climate for future generations" [150].

The link between the sustainable development and public health is obvious and does not have to be emphasized. [59] introduces direct and indirect routes by which energy sources may affect human health. This complex system can be decomposed into sub-systems:

- Air pollutants
- Climate change
- Water

- Ecological systems
- Social/economic systems

Each of the sub-systems affects human health. The strength and dynamics of health outcomes can be measured with statistical indicators: mortality, morbidity, birth defects rate, etc. The "Health" concept represents a complex system, which includes physical, social, mental, spiritual and biological well-being, spanning across all the spheres of human lives. Environmental pollution, as one of the factors with dominant and obvious influence upon human health, causes direct and latent harmful effects, which must be evaluated in order to create a set of preventive health-preserving solutions. That is why linking all the named components in one system and studying of this system leads to the analysis of potential and present health problems, retrieval of the new ones and to working out the in-depth view of situation development, strategies, and activities oriented to situation management and control.

As it was noted in the European Scientific Committee's issues [155], the data processing of data for human impact assessment "has turned out to be very laborious and resource-demanding", because "there is a limited access to data on environmental exposures and health status in specific geographical areas". Numerous studies have shown an adverse relationship between environmental hazards and human health. Some research works were aimed to discover detail mechanisms of these relationships [153], [81], [16], [178], [198]. For example, a known and aggressive air contaminant is ambient fine particulate matter ($PM_{2.5}$). It has been demonstrated that it increases cardiovascular risks, with a stronger impact on heart failure [72] and causes premature death if it is locally emitted [137]. Stieb [190] has examined and proved the presence of associations between carbon monoxide (CO), nitrogen dioxide (NO_2), ozone (O_3), sulfur dioxide (SO_2), and particulate matter (PM_{10} and $PM_{2.5}$), and visits for angina/myocardial infarction, heart failure, dysrhythmia/conduction disturbance, asthma, chronic obstructive pulmonary disease, and respiratory infections.

White et. al. [202] describe the research in which adverse respiratory health effects on children caused by the petrochemical refinery's emissions are studied. The emissions include SO_2, particles and oxides of nitrogen as well as fugitive emissions consisting of numerous aliphatic and aromatic hydrocarbons were studied. Some studies of prenatal exposure to solvents including tetrachloroethylene have shown increases in the risk of certain congenital anomalies among exposed offspring [3]. Some of the environmental pollutants play the role of carcinogens. For example, there are two types of environmental exposures, which are related to lung cancer: radon in homes and arsenic in drinking water [21]. Migliore et. al. [129] have studied a causal effect that leads to increased respiratory illnesses in children due to traffic pollution.

4.2.2 Environmental Impact Assessment

Unfortunately, decision making in environmental health is not a simple task. Indeed, retrospective environmental data are skewed by noise, gaps and outliers, as well as measurement errors. Working with public health information adds restrictions caused by the methodologies of data measurement, the standards currently in use, data availability, and so on. For example, it is known that the International Statistical Classification of Diseases and Related Health Problems (ICD) was reviewed 10 times, the International Classification of Functioning and Disability (ICIDH) another 2 times, also, any local standard is also reviewed [83]. In recent years, it has been proven that it is essential to use products and energy life cycle indicators in order to assess the ecological impact. Some difficulties in obtaining this information have been noticed. The idea was developed and fixed in International Standard ISO 14031 "Environmental Management - Environmental Performance Evaluation - Guidelines", which certifies the usage of indirect indicators [88].

The multi-agent approach is an excellent technique that can help to reduce the complexity of a system by creating modular components, which solve private subtasks that together achieve the whole goal. Every agent utilizes the most effective technique for solving its subtask and does not apply the general approach, which is often acceptable for the system as a whole, but not optimal for a concrete subtask [185].

In case of environmental impact assessment (EIA), all the advantages of intelligent agents become crucial. EIA is an indicator, which enables evaluation of the negative impact upon human health caused by environmental pollution. This kind of pollution, a factor with dominant and obvious influence, causes direct and latent harm, which must be evaluated and simulated in order to create a set of preventive health-preserving solutions. Large amounts of raw data describe the "Environment-Human health" system, but not all the information is used. The system transforms from the initial "raw" state to the "information" state, which suggests organized data sets, models and dependencies, and, finally, to the "new information" which is represented as a set of recommendations, risk assessment and forecast values.

4.3 Design of the Agent-Based Decision Support System

4.3.1 Meta-ontology of the System

Meta-ontology consists of five private ontologies and includes the following models: domain of interest, aims and tasks, agents, interaction, and environment (see Section 3.5).

1. Ontology of Environment
2. Ontology of MAS Architecture
3. Ontology of Agents

4. Ontology of Interactions
5. Ontology of Tasks

The Protégé Ontology Editor was used to design and implement these five ontologies [151], [179]. Initially, the Ontology of Environment was created. For the current case of the study, the Ontology of Environment is able to describe the hierarchy of the "Environment-Human health" system, which includes the following entities:

- To describe the "human health" concept:

  - Morbidity by classes of diseases or external reasons
  - Endogenous and exogenous morbidity

- To describe the "environmental impact" concept:

  - Air pollutants
  - Pollutants of soil
  - Potable water pollutants
  - Noise contamination

These entities are arranged into classes. Each class has its properties, values, restrictions and axiomatic rules. Using the Protégé Ontology Editor, properties were divided into "object properties" and "data properties". Fig. 4.1 gives a view of the main classes that were created for the meta-ontology, which shows an asserted hierarchy of the classes. It contains classes for the private ontologies, which are disjoint of each other. The Ontology of Environment contains individuals from two classes: the *individuals_pollution* and *individuals_morbidity* subclasses. These classes were created because "morbidity" and "pollution" individuals have different properties.

In order to show an example of the way how an entity of the *Classes* is represented, the general description of the domain of interest is introduced. It includes regions, which are characterized with some environmental pollution and human health level. The ontology for the *Morbidity* class is represented in Fig. 4.2 and described as:

$$Morbidity = M^t = < m_j, m_{j,k}^{t,g,ag} >$$

where m is a morbidity class from the M, which is the set of morbidity classes, $m \in M$, t represents the time of registration (year), $j = 1 \ldots |M|$, k stands for a general class of disease (endogenous or exogenous) $k = 1, 2$, g is the gender, and ag stands for the age. The superclass-subclass relations are stated through the indexes, as shown for the class *Morbidity* in Fig. 4.2.

For the *Pollution* class, its formal description is:

$$Pollution = P_t = < p_i, p_{ij}, p_{ij}^t >,$$

where p represents a pollutant from the set of main pollutants P, $p \in P$, p_i is a sub-pollutant from class p, t is the time of registration (year), $i = 1 \ldots |P|$, $j = 1 \ldots |P_i|$.

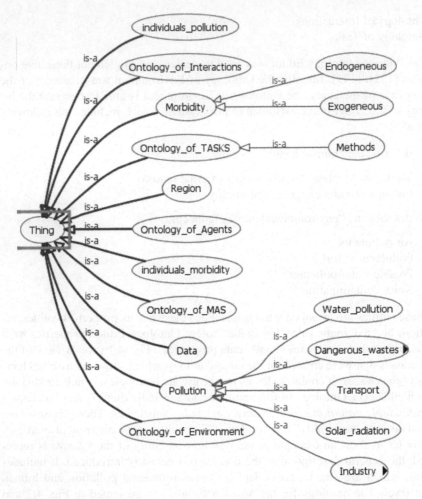

Fig. 4.1 The general view on the classes of the meta-ontology realized in Protégé 4.0

The *Pollution* class is represented in a similar manner as the class *Morbidity*, as it shown in Fig.4.3.

As stated previously, the *Morbidity* class includes two subclasses of diseases: endogenous and exogenous, which are detailed into classes in accordance with the International Classification of Diseases [83]. The *Environment* class includes the following performance indicators: water pollution, dangerous wastes, transport activity, and industrial activity parameters revealing dangerous emissions during energy life-cycles (use of energy, and so on). Fig. 4.4 illustrates the individual, which belongs to the *individuals_morbidity* class. The individual is named "A1992EN2F".

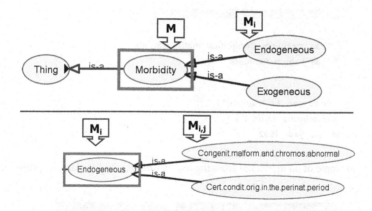

Fig. 4.2 Illustration of *Morbidity* class representation

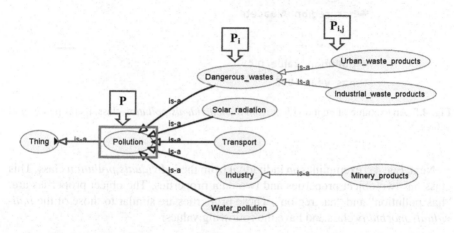

Fig. 4.3 Illustration of *Pollution* class representation

This class has two object properties: "has_morbidity" and "has_region". These properties contain other objects as values. The first one contains the "Congenital malformations" individual from the *Morbidity* class as a value, and the last one contains the "Albacete" value from the *Region* class. The three other properties belong to "data property" type, which may contain values from simple types: Strings, integer, double values, enumerations, etc. For the individual "A1992EN2F" the properties have the following values:

- "has_gender" with the String value "Female" from enumeration list *Gender*,
- "has_year" with the value "1992" from the enumeration list *Year*, and
- "morbidity_value" with the double value "2.2".

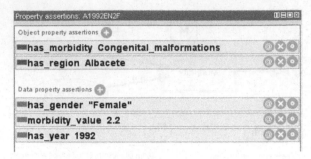

Fig. 4.4 An example of an individual from the *individuals_morbidity* class, its properties and values

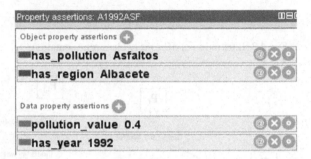

Fig. 4.5 An example of an individual from the *individuals_pollution* class, its properties and values

Next, Fig. 4.5 exemplifies an individual from the *individuals_pollution* class. This class has two object properties and two data properties. The object properties are: "has_pollution" and "has_region". These properties are similar to those of the *individuals_morbidity* class and have the following values:

- "has_pollution" has a reference to an object "Asphalts". "Asphalts" is a member of the *Industry* subclass of the *Pollution* class.
- "has_region" contains the "Albacete" value from the *Region* class.

The data properties contain the following values:

- "has_year" with the value "1992" from the enumeration list *Year*, and
- "pollution_value" with the double value "0.4".

An individual "DataFusion" of the Ontology of MAS Architecture is shown in Fig. 4.6. It´s "Level", "Role" and "Order" components are represented with the following data properties:

- "Level_value" that contains a value of integer type,
- "role_ID" that contains an ID of the individual, and
- "order" that contains a value of integer type.

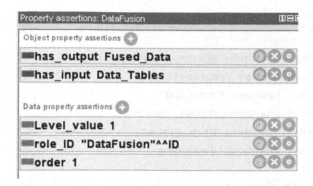

Fig. 4.6 An example of an individual from the Ontology of MAS Architecture

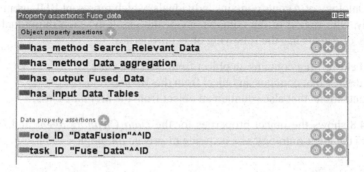

Fig. 4.7 An example of an individual from the Ontology of Tasks

The "InformationFlows" component of the Ontology of MAS Architecture is represented with two object properties: "has_output" and "has_input", which link the individual with the *Data* class. The first property sets up the input data, and the last one sets up the output data of the role.

An individual "Fuse_data" of the Ontology of Tasks is shown in Fig. 4.7. The correspondence between the components of the ontology and individual properties is as follows:

- the component "Task" is represented with the data property "task_ID",
- the component "Method" is represented with the object property "has_method",
- the components "Input" and "Output" are represented with the object properties "has_output" and "has_input", which contain objects from the *Data* class, and
- the component "Role" is represented with the data property "role_ID" that contains ID of the correspondent role.

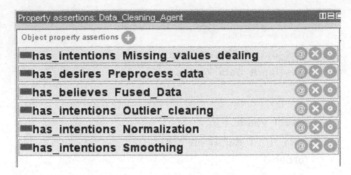

Fig. 4.8 An example of an individual from the Ontology of Agents

The Ontology of Agents contains individuals, which represent BDI agents. Consequently, with regard to the components of the ontology, each individual of the Ontology of Agents has the following object properties:

- "has_believe" that contains an object from the Data class,
- "has_desire" that contains an object from the *Methods* class,
- "has_intention" that alsos contain an object from the *Methods* class,

Fig. 4.8 shows the object properties for the *Data_Cleaning_agent* and Fig. 4.9 illustrates the meta-ontology as a mapping of private ontologies.

4.3.2   Logical Levels of the ADSS

The DeciMaS framework consists of three phases, which are reflected in the architecture of the ADSS. The proposed system is logically and functionally divided into three layers: the first is dedicated to meta-data creation (information fusion), the second is aimed at knowledge discovery (data mining), and the third layer provides real-time generation of alternative scenarios for decision making [180]. The levels do not have strongly fixed boundaries, because the agents construct a community, in which the agents´ spheres of competence can overlap, and the boundaries are smooth. The ADSS asserts the main points of a traditional decision making process, and includes the following steps:

1. Problem definition
2. Information gathering
3. Alternative actions identification
4. Alternatives evaluation
5. Selection of decision
6. Decision implementation

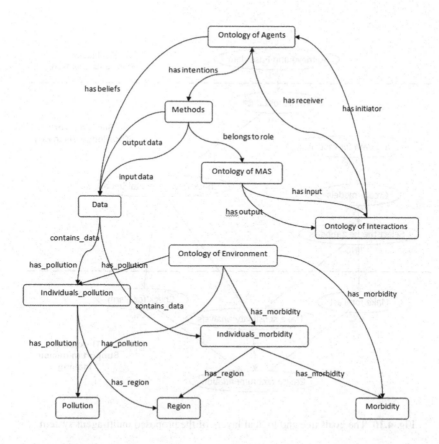

Fig. 4.9 Meta-ontology as a result of private ontologies mapping

The first and the second stages are performed during the initial step, when the expert information and initial retrospective data are gathered, stages three, four and five are solved by means of the MAS, and the sixth stage is supposed to be carried out by the decision maker. Although the goals of the ADSS are determined and kept constant within various domains, the goals of the case study are not visible and clear at first sight. That is why the domain of interest has been studied with the creation of a goal tree. Prometheus Design Tool (PDT) offers graphical tools to design the ADSS.

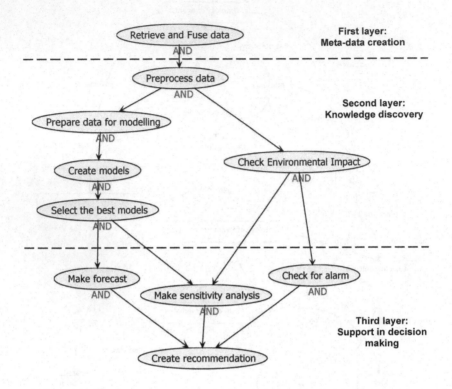

Fig. 4.10 The goals tree and logical layers of the proposed multi-agent system

4.3.3 The Principal Abstractions of the System

4.3.3.1 Goals and Scenarios

The principal goals of the proposed multi-agent system follow the logical sequence of the main stages of the DeciMaS framework. For these reasons, the process of the system design starts with identification of general goals that are divided into subgoals and then refined. Fig. 4.10 shows the goals tree specified for the system, and where the logical systems' layers are marked.

The final goal is *Create recommendation* and it is achieved as a result of three parallel goals *Make forecast*, *Make sensitivity analysis* and *Check for alarm*. The goals *Make forecast* and *Make sensitivity analysis* use parts of the same knowledge. Both sensitivity analysis and forecasting are based on models received as a result of *Create models* and *Select the best models* goals. The goal *Check environmental impact* is independent from the other goals of the second logical layer, although its

Table 4.1 The motivational scenario for achieving the "Retrieve and fuse data" goal

Name	**Retrieve and fuse data** scenario				
Actors	Expert				
Initiator	The Data Aggregation agent				
Trigger	The "Read ontology" message to the Domain Ontology agent				
Step #	Type	Name	Role	Data used	Data created
1	Percept	Obtain expert knowledge	Data Fusion	Domain Ontology agent beliefs	Domain Ontology agent beliefs
2	Action	Read ontology and data sources	Data Fusion	External data sources	Domain Ontology agent beliefs
3	Action	Fuse morbidity data	Data Fusion	External data sources	Domain Ontology agent beliefs
4	Action	Fuse transport data	Data Fusion	External data sources	Domain Ontology agent beliefs
5	Action	Fuse data on mines	Data Fusion	External data sources	Domain Ontology agent beliefs
6	Action	Fuse petroleum usage data	Data Fusion	External data sources	Domain Ontology agent beliefs
7	Action	Fuse data on wastes	Data Fusion	External data sources	Domain Ontology agent beliefs

outcomes are used for making recommendations on the third logical layer. The goal *Preprocess data* is the initial goal for all the data mining procedures. This fact is illustrated (see Fig. 4.10) as the goal *Preprocess data* inherits meta-data from the first layer and from the goal *Retrieve and Fuse data*.

Scenario is one of the key concepts in Prometheus methodology, as it represents a purposeful behavioral model of collective activity [189]. A scenario serves to achieve a practical goal and may include sub-scenarios as well. It involves, at least, two agents performing particular roles. A scenario is a composition of agents' scenarios oriented to achieve goals, which in their turn, may be achieved by sets of problem-solving activities that include agent roles. Every role played by an agent includes actions and plans. Table 4.1 to Table 4.5 present five scenarios that the system should enact. Each scenario is described as a sequence of steps, where each step has its type, name, role, and type of the data it uses and produces. The type term can be a goal, an action, a percept, or a sub-scenario.

Table 4.2 The motivational scenario for achieving the "Preprocess data" goal

Name	**Preprocess data** scenario				
Actors	No				
Initiator	The Data Preprocessing agent				
Trigger	The "ready" message from the Data Aggregation agent				
Step #	Type	Name	Role	Data used	Data created
1	Action	Eliminate arti-facts	Data Clearing	Domain Ontology agent beliefs	Morbidity, Pollu-tants
2	Action	Normalize data	Data Clearing	External data sources	Morbidity, Pollu-tants
3	Action	Smooth data	Data Clearing	Morbidity, Pollu-tants	Morbidity, Pollu-tants
4	Goal	Prepare data for modeling	Data Clearing	Morbidity, Pollu-tants	dataX, dataY
5	Action	Parametric corre-lation	Data Clearing	dataX, dataY	Correlation table
6	Action	Non-parametric correlation	Data Clearing	dataX, dataY	Correlation table

The scenario of achieving a ***Retrieve and fuse data*** goal is given in Table 4.1. It is initiated by the Data Aggregation agent, and the Domain ontology and the data mining agents participate in this scenario. It suggests collaboration with the actor EXPERT. The scenario has seven steps. The first step supposes receiving the perception "Obtain expert knowledge". The remaining steps are actions. The scenario to achieve a ***Preprocess data*** goal is given in Table 4.2 and it describes interaction of the Data Aggregation agent and the Data Preprocessing agent and its team. This scenario has one goal ***Prepare_data_for_modeling*** and performs five actions, namely *Eliminate artifacts*, *Normalize data*, *Smooth data*, *Parametric correlation* and *Non-parametric correlation*.

The **Check Environmental Impact** scenario has two actions: *Create neural networks* and *Evaluate impact assessment*, and invokes the Function approximation agent and the Artificial Neural Network agent. This scenario is presented in Table 4.3. The **Create models** scenario has eleven steps, which are actions to be completed by agents from the Function Approximation agent team, as it is shown in Table 4.4. The **Create recommendation** scenario suggests collaboration with the external actor USER/DECISION MAKER and contains ten steps. These steps are carried out within two roles: *Computer simulation* and *Decision making*. Table 4.5 provides a detailed view of the scenario containing these roles.

Table 4.3 The motivational scenario for achieving the "Check Environmental Impact" goal

Name	**Check Environmental Impact** scenario				
Actors	No				
Initiator	The Function Approximation agent				
Trigger	The "Start Data Mining" message from the Function Approximation agent				

Step #	Type	Name	Role	Data used	Data created
1	Action	Create neural Networks	Impact Assessment	dataX, dataY, Correlation table	IAResults
2	Action	Evaluate Impact Assessment	Impact Assessment	IAResults	IAResults

4.3.3.2 Actors

Being implemented by means of the Prometheus Design Tool, the Analysis
Overview Diagram of the MAS illustrates the high-level view composed of external
actors, key scenarios and actions, as shown in Fig. 4.11. The proposed ADSS supposes communication with two actors. One actor, EXPERT, embodies the external
entity which possesses the information about the problem area. In more detail, the
EXPERT contains the knowledge of the domain of interest represented as an ontology, and delivers the knowledge through protocol *ReturnEI* to the ADSS.

The data source "The CS Results", stores the results of the simulation and forms
a knowledge base. Through the **Simulate Models** scenario, the user interacts
with the knowledge base, and gets recommendations if they have been previously
simulated and stored before, or creates and simulates the new alternative decisions.
As a result of the interaction within the **Retrieve and Fuse data** scenario,
the raw information is read, and then shown as "Heterogeneous Data Sources" data
storage. The "Pollutants" and "Morbidity" data sources are created afterwards. The
second actor, named USER/DECISION MAKER, is involved in an interactive process
of generation of alternative decisions to choose the optimal one/ones. This actor
communicates with the agents by passing a message through protocol *ReturnSUI*,
stating the model, simulating values, predicting periods, levels of variable change,
etc. The actor accepts the best alternative in accordance with its beliefs and the
MAS.

The flow of works, which are essential for decision making, include three sub-
scenarios: the **Simulate models** scenario, the **Create recommendation**
scenario and the **Search for the adequate** model scenario. Additionally,
there are three goals, which are related to each scenario and have similar names.
Each goal has a number of activities, and within each scenario resources in the form
of data sources are used, modified or created.

Table 4.4 The motivational scenario for achieving the "Create models" goal

Name	**Create models** scenario				
Actors	No				
Initiator	The Function Approximation agent				
Trigger	The "ready" message from the Data Preprocessing agent				

Step #	Type	Name	Role	Data used	Data created
1	Action	Decomposition	Decomposition	dataX, dataY, Correlation table, ranks	Groupings table
2	Action	Create univariate regression models	Function Approximation	dataX, dataY, Correlation table	Models table
3	Action	Create multiple regression models	Function Approximation	dataX, dataY, Correlation table	Models table
4	Action	Create neural network models	Function Approximation	dataX, dataY, Correlation table	Models table
5	Action	Create GMDH-models	Function Approximation	Models table, dataX, dataY	Models table
6	Action	Evaluate univariate regression models	Function Approximation	Models table	Models table
7	Action	Evaluate multiple regression models	Function Approximation	Models table	Models table
8	Action	Evaluate neural network models	Function Approximation	Models table	Models table
9	Action	Accept models	Function Approximation	Models table	Models table
10	Action	Create committee machines	Function Approximation, Impact Assessment	Models table	Final models
11	Action	Creation of reports	Function Approximation, Impact Assessment, Decomposition	IAResults, Final models, CS Results, Grouping table	

4.3.3.3 Roles of the Proposed MAS

Scenarios focus on how a multi-agent system achieves goals, interaction models define the agent´s outgoing communications, and the actors represent external entities, which interact with the system [189]. The detailed behavior of an agent is

Table 4.5 The motivational scenario for achieving the "Create recommendation" goal

Name	**Create recommendation** scenario
Actors	User / Decision maker
Initiator	The Computer Simulation agent
Trigger	The "ready" message from the Function Approximation agent

Step #	Type	Name	Role	Data used	Data created
1	Percept	Obtain preferences for simulation	Computer Simulation		Data for simulation
2	Goal	Make forecast			
3	Action	Forecasting	Computer Simulation	Models tables, dataX, dataY	CS Results table
4	Goal	Make sensitivity analysis	Computer Simulation	dataX, dataY, Correlation table	CS Results table
5	Action	Models simulation	Computer Simulation	Models table, dataX, dataY	CS Results table
6	Percept	Preferences for decision	Decision Making		Data for decision
7	Action	Criteria application	Decision Making	CS Results table, IAResults	Final models
8	Goal	Check for alarm	Decision Making		
9	Action	Alarm generation	Decision Making	Alarm levels	Final models
10	Action	Creation of reports	Decision Making	IAResults, Final models, CS Results	

represented by roles and plans. The communication between agents is shown in acquaintance models. Roles represent agent´s functions, responsibilities and expectations. A role enables pooling together the goals of the system in accordance with different types of behavior that an agent assumes when archiving a goal or a series of goals. Table 4.6 shows a view of the correspondent logical levels, and the roles, which are played there.

The distribution of roles for agents determines their specialization and knowledge (see Fig. 4.12). One of the intentions for the system design was to assign one role to each agent or agent team. That requirement was met for the roles *Data Fusion* and *Data Clearing* where the teams of the Data Fusion agent and the Data Preprocessing agent carry out these roles. Moreover, the Function Approximation agent manages three data mining roles: *Impact Assessment*, *Decomposition* and *Function Approximation*, and the Computer Simulation agent takes on *Computer Simulation*, *Decision Making* and *Data Distribution* roles.

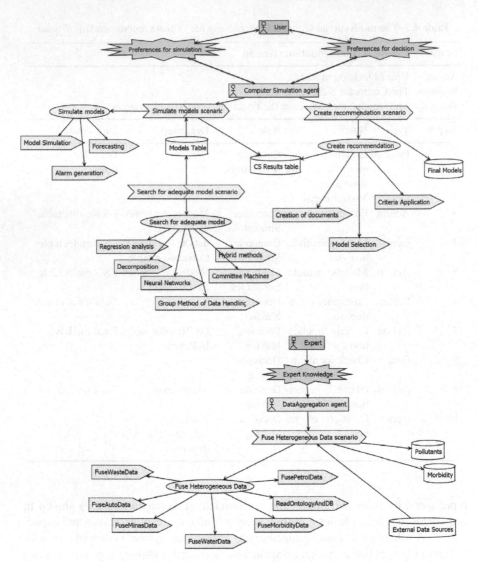

Fig. 4.11 The Prometheus Diagram of MAS interaction with actors

4.3.3.4 Data and Interactions within the System

During the system work cycle, agents manipulate diverse income and outcome information flows: data transmission protocols, messages, income and outcome data, etc. These information sources have different times of usage. They can be permanent or can be created to support a specific task and then be destroyed. They can also be classified by the levels of access: data may have public, agent or private access.

Table 4.6 The roles played in the MAS

Logical level	Main agent	Subordinate agent	Role
Data Fusion	Data Aggregation agent	Domain Ontology agent Traffic Pollution Fusion agent Water Data Fusion agent Petroleum Data Fusion agent Mining Data Fusion agent Morbidity Data Fusion agent Waste Data Fusion agent	Data Fusion
	Data Preprocessing agent	Normalization agent Correlation agent Data Smoothing agent Gaps and Artifacts Check agent	Data Clearing
Data Mining	Function Approximation agent	Regression agent ANN agent GMDH agent Committee Machine agent Decomposition agent Evaluation agent	Impact Assessment Decomposition Function Approximation
Decision Making	Computer Simulation agent	Forecasting agent View agent Alarm agent	Computer Simulation Decision Making Data Distribution

The description of the data sources, which are created and used by the agents, is illustrated in Fig. 4.13 as a Prometheus "Data coupling" diagram.

In Fig. 4.13, the "External Data Sources" are data sources that contain information of interest. "DomainOntologyAgent_private_data" is data that contains information about the Domain of Interest Ontology. This data is used within the *Data Fusion* role. The "Morbidity" and "Pollution" data storages are created by agents that act within the *Data Fusion* role. These data contain fused and homogenized information. The arrows that coming from "Morbidity" and "Pollution" data storages and pointing to the *Data Cleaning* role indicate that they are used within this role.

The "dataX", "dataY", "ranks" and "Correlation Table" data are created during the "Data Cleaning" role. The "dataX" contains cleared and normalized data for the independent variables **X** that represent pollutants. The "dataY" contains cleared and normalized data for the dependent variables **Y** that represent morbidity classes. The "ranks" and "Correlation Table" data are also created during the *Data Cleaning*

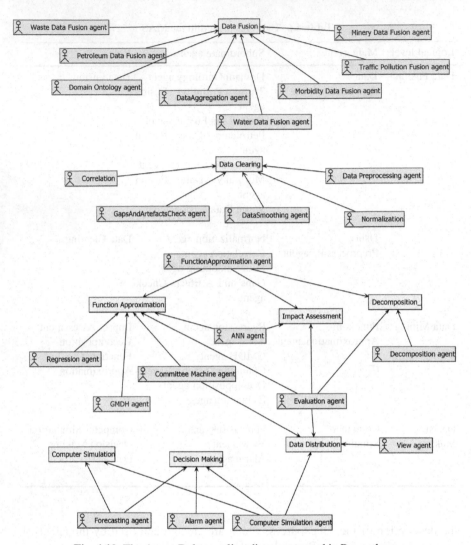

Fig. 4.12 The Agent-Role coupling diagram created in Prometheus

execution of the role. They contain results of correlation analysis. The role *Decomposition* uses "dataX", "dataY" and "ranks" data sources to carry out decomposition and creates the "Groupings table" data. The "Models table" data is created during the execution of the *Function Approximation* role. It contains data about approximated models. The role *Impact Assessment* uses "dataX", "dataY", and "Correlation Table" data sources and creates "IA Results" data, which contains outcomes of environmental impact assessment. The role *Computer Simulation* requires "Models table" and "IA Results" data to calculate alternative decisions and stores them into the knowledge base "CS Results List". The role *Decision making* supposes interaction with a user. The user makes his/her preferences and selects alternatives from

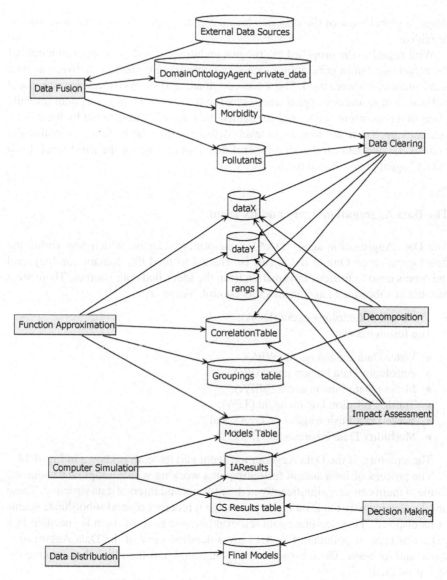

Fig. 4.13 Data and roles coupling diagram in Prometheus

the "CS Results List", which are moved to the "Final Models" data. During the execution of the *Data Distribution* role, "IA Results" and "Final Models" data sources are used.

4.3.3.5 Description of the Agents

Once the multi-agent system notions have been defined and its logical architecture has been determined, with the set of goals, scenarios, interaction models and data usage, a global view of the system´s layers and description of agent teams may be provided.

With regard to the proposed multi-agent architecture and in order to gain time of the recommendation generation process and optimize interactions between agents, local agent teams were used. The teams coordinate and supervise task execution and utilization of resources. Agent teams synchronize the work of the system, execute plans in a concurrent mode, and strengthen the internal management by local decision making. There are four agent teams defined within the system: two within the first level, one team on the second level and another one on the third level. Each "main" agent plays several roles.

The Data Aggregation Agent and Its Team

The Data Aggregation agent (DAA) is the principal agent, which acts within the first logical layer. One of the agents is oriented to read the domain ontology, and the others have to retrieve information from the identified data sources. There are a number of subordinate agents under its control. These are:

1. The Domain Ontology agent (DOA).
2. The fusion agents:

   • Water Data Fusion agent (WFA),
   • Petroleum Data Fusion agent (PFA),
   • Mining Data Fusion agent (MiFA),
   • Traffic Pollution Fusion agent (TFA),
   • Waste Data Fusion agent (WDFA), and,
   • Morbidity Data Fusion agent (MoFA).

The structure of the Data Aggregation agent and its team is shown in Fig. 4.14.

The process of information fusion requires working with multiple data sources. Some of them can vary significantly in their format and internal data structure. These are the reasons why the Data Aggregation agent receives several subordinate agents at its disposal. They facilitate data retrieval because each of them is specified in a particular type of pollutant. Fig. 4.15 gives detailed view of the Data Aggregation agent and its team. The messages that agents send and the plans they execute are also presented.

The Data Aggregation agent starts working with sending the message *ReadOntology* to the Domain Ontology agent, which reads the OWL-file containing information about the ontology of the domain. Then the Domain Ontology agent passes this information to the Data Aggregation agent. The Domain Ontology agent completes its execution by sending the message *OntologyIsBeingRead* to the Data Aggregation agent. Next, the Data Aggregation agent sends a *Start Fusion* message to the fusion agents, which initiate execution. When beginning execution, each fusion agent searches for the files that may contain information about the concept of its interest. Each fusion agent works with one or several concepts of the domain ontology. The Water Data Fusion agent searches for the information about water

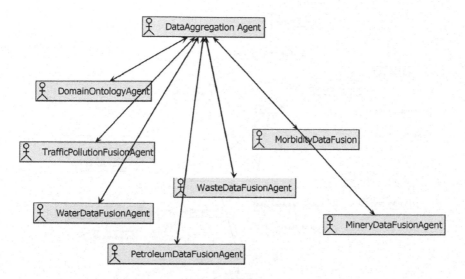

Fig. 4.14 The Data Aggregation agent and its team

contaminants and their properties. The Petroleum Data Fusion agent retrieves information about the use of petroleum and related concepts. The Mining Data Fusion agent retrieves data about the contamination related to mining industry activity. The Waste Data Fusion agent retrieves data about wastes and its components. The Transport Data Fusion agent extracts data about transport vehicles activity. And, finally, the Morbidity Data Fusion agent searches for the information about morbidity and its properties.

When an agent finds the information file, it starts to retrieve information about the concept and its values. It then changes their properties (in order to eliminate heterogeneity and to homogenize information) and sends it to the Data Aggregation agent, which pools the retrieved information together. Finally, Data Aggregation agent fills the domain ontology with data, and puts that data into a standard format. Next, the data files are ready to be preprocessed, and the Data Aggregation agent sends the message *Fusion Is Finished* to the Data Preprocessing agent through the protocol *ReturnDF*. This means that the data are fused and preprocessing can begin. The message is generated only when all the fusion agents have finished their execution (see Fig. 4.15). In the current case study, information is weakly organized, and is presented in the form of plain text files, tables or consolidated forms. In this case, it is necessary to analyze the file structure, and localize the principal concepts and their properties, which can be found as intersections of the concepts or the concepts and their properties. Thereby, data extraction turns into file content analysis.

The Data Aggregation agent must achieve the following goals:

1. Obtain information from the ontology of the domain.
2. Search for information sources, which may contain information of interest stored in the ontology of the domain.

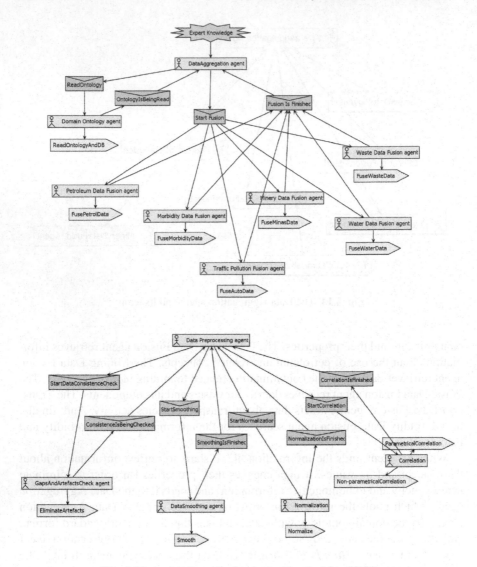

Fig. 4.15 Agents from the first logical level of the ADSS

3. Retrieve information from the found sources.
4. Transform the retrieved information in order to avoid heterogeneity.
5. Fuse information.

The Data Aggregation agent has the internal databases *OntologyDB* and *DOA-_privatedata*. The *OntologyDB* contains concepts from the Ontology of the Domain of Interest. Each of the data fusion agents receive their part of knowledge from this internal database. The Data Aggregation agent interacts with the EXPERT actor and receives information from it. This information is shown as perception "Expert

Input: Data Sources (**DS**), which may contain concepts $C=\{C_1, C_2, ..., C_n\}$ and properties $P=\{P1, P2, .., Pm\}$.
Output: Retrieved information **C** and **P**.

(1) Search for available data sources and for each of them:
    a. Check the type of **DS** and call for the appropriate **FA** agent.
    b. Wait until the **FA** agents terminate its execution.
    c. Receive retrieved information from each **FA** agent.
    d. Paste retrieved concepts **C** and **P**.
(2) Form meta-ontology.
(3) Send a message to the **DPA**.

Fig. 4.16 The algorithm for the Data Aggregation agent functionality

knowledge". The algorithm of the functionality for the Data Aggregation agent is shown in Fig. 4.16, where DS stands for data source, FA for a fusion agent, and DPA for the Data Preprocessing agent.

The Domain Ontology agent is a member of the Data Aggregation agent team, and is triggered by the message *ReadOntology* (see Fig. 4.15). It begins execution in agreement with the following algorithm. First, it searches for the *owl-file*, which contains information about the ontology of domain. The Domain Ontology agent has to assess the OWL-file, copy it and pass it to the Data Aggregation agent. The internal structure of the Domain Ontology agent is shown in Fig. 4.17.

The Domain Ontology agent aims to obtain knowledge about the Ontology of Domain of Interest from the OWL-file, and pass it to the Data Aggregation agent. As a result, the hierarchy of concepts with properties is created. To achieve its goal, the Domain Ontology agent has a plan, which is named *readOntologyAndDB*. When the **ReadOntology** event occurs, the Domain Ontology agent begins execution of the plan (see Fig. 4.17). The Domain Ontology agent parses the OWL-file in order to retrieve concepts to form the data storage in accordance with the statement 3.2. When the agent has been executed, it sends the *OntologyIsBeingRead* message to the Data Aggregation agent.

The general view of the interaction of the Domain Ontology agent and the Data Aggregation agent with the Ontology of Domain of Interest is illustrated in Fig. 4.18. Fig. 4.18 gives a view of the meta-ontology, where the Ontology of Domain is highlighted.

The Data Preprocessing Agent and Its Team

The Data Preprocessing agent (DPA) aims to prepare the initial data for further modeling. It manages a number of subordinate agents, which make up its team. Each subordinate agent specializes in a different data clearing technique:

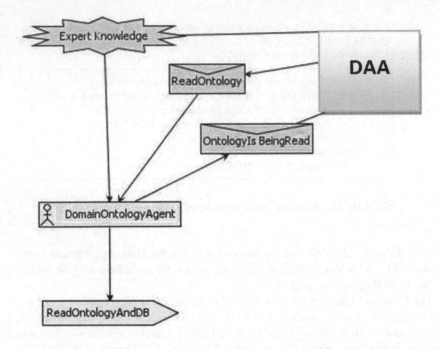

Fig. 4.17 Internal structure and reasoning of the Domain Ontology agent

- Gaps and Artifacts Check agent (GAA) clears fused raw information from miss-
  ing and inconsistent values and fills the gaps
- Data Smoothing agent (DSA) carries out exponential and moving average
  smoothing procedures
- Normalization agent (NA) normalizes data sets
- Correlation agent (CA) calculates correlation matrices

Fig. 4.19 is a part of Prometheus agent acquaintance diagram, and it provides a
view of the Data Preprocessing agent and its team. Data transformation tasks cannot
be executed in a concurrent mode, because the output of the previous data clearing
procedure is the input to the next one. In fact, the main function of the Data Preprocessing agent is to coordinate the subordinate agents and to decide when they
execute and in which order. The algorithm of the Data Preprocessing agent working
cycle is given in Fig. 4.20).

The Data Preprocessing agent starts to execute as soon as it receives a triggering message from the Data Aggregation agent (see Fig. 4.21). In the beginning of
its execution, the Data Preprocessing agent sends the *StartDataConsistenceCheck*
message to trigger the Gaps and Artifacts Check agent, which eliminates artifacts,
searches for the double values, and fills the gaps. Having finished execution, it sends
a message to the Data Preprocessing agent.

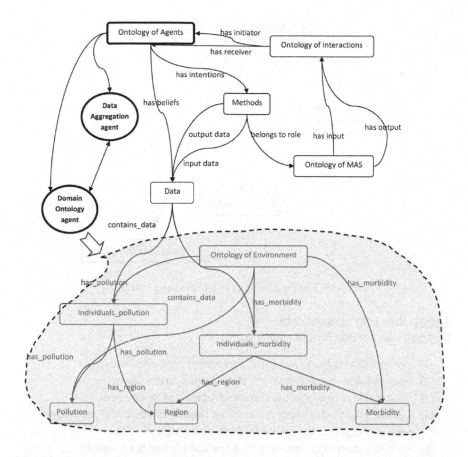

Fig. 4.18 The Data Aggregation agent, the Domain Ontology agent, and the meta-ontology: the agents work with the Ontology of Environment. The dotted connector between the Data Aggregation agent, the Domain Ontology agent and the Ontology of Agent shows that the agents are objects of the mentioned ontology. The double arrows indicate that there are input and output information flows

The next data clearing task is smoothing, and the Data Preprocessing agent calls for the Data Smoothing agent by sending the message *StartSmoothing*. The Data Smoothing agent can execute exponential and weighted-average smoothing. It completes execution by sending the *SmoothingIsFinished* message to the Data Preprocessing agent. Next, the Data Preprocessing agent sends a message to trigger the Normalization agent, which executes several normalization algorithms. The Data Preprocessing agent creates data sources, which contain normalized data. These sources are used later in the second logical level for modeling. Lastly, the Data Preprocessing agent triggers the Correlation agent. It calculates correlation matrices for preprocessed data. The outputs of the Data Preprocessing agent work include:

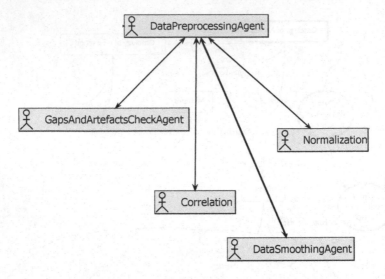

Fig. 4.19 The Data Preprocessing agent and its team

Input: Raw unprocessed data.
Output: Data ready for data mining.

(1) On start fill in believes with initial information from meta-ontology.
(2) Read the start message from the **Data Aggregation agent**.
(3) Send message to trigger the **GapsAndArtefactsCheck agent**.
(4) Receive a termination message from the **GapsAndArtefactsCheck agent**.
(5) Send message to trigger the **DataSmoothing agent**.
(6) Receive a termination message from the **DataSmoothing agent**.
(7) Send message to trigger the **Normalization agent**.
(8) Receive a termination message from the **Normalization agent**.
(9) Send message to trigger the **Correlation agent**.
(10) Receive a termination message from the **Correlation agent**.
(11) Send a message to the **Function Approximation agent**.

Fig. 4.20 The working algorithm of the Data Preprocessing agent

data (ready for further processing and modeling) and additional data sources with correlation and normalization results.

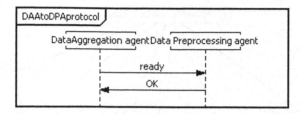

Fig. 4.21 The triggering message from the Data Aggregation agent to the Data Preprocessing agent

The Function Approximation Agent and Its Team

The Function Approximation agent (FAA) has a hierarchical team of subordinate agents, which serve to carry out the roles: "Impact Assessment", "Decomposition" and "Function Approximation". FAA has under its control a number of subordinate agents:

- Data mining agents, which work in a concurrent mode and create models of the following types:

  - The Regression agent (RA),which creates regression models.
  - The ANN agent (AA), which creates neural network models.
  - The GMDH agent (GMDHA), which creates polynomial models with the group method of data handling.

- The Evaluation agent (EA), that calculates evaluation criteria for models.
- The Committee Machine agent (CMA) that creates hybrid models.
- The Decomposition agent (DA) that carries out the decomposition procedure.

Fig. 4.22 shows the Function Approximation agent and its team of agents, whereas Fig. 4.23 shows the Function Approximation agent and its team in detail. The agents communicate by sending messages. Also, the plans of the agents are shown.

The Function Approximation agent manages subordinate agents. The FAA working cycle is as follows. The Function Approximation agent sends the *Start-Decomposition* message and waits until the DA finishes its execution. After decomposition, the Function Approximation agent has the necessary information for modeling. Therefore, it sends the *StartDataMining* message to each of the data mining agents. They begin execution in a concurrent mode. When any agent from this group finishes modeling, it sends the *StartEvaluation* message to the Evaluation agent.

The Evaluation agent evaluates the received models and checks the adequacy of the model against the experimental data, and returns the list of the accepted models, while the others are banned and deleted. The Evaluation agent communicates that its execution is completed by sending the *EvaluationIsFinished* message to a data

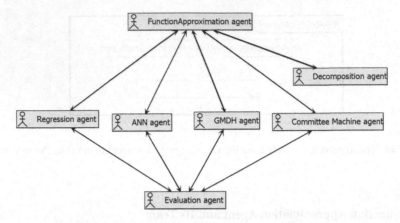

Fig. 4.22 Function Approximation agent and its team

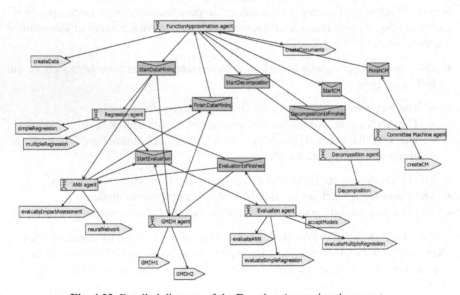

Fig. 4.23 Detailed diagram of the Function Approximation agent

mining agent. In turn, the data mining agent generates the *FinishDataMining* message. When all the data mining agents finish their execution, they send messages to the Function Approximation agent, which pools the output of the agents work and creates the list of accepted models.

Finally, the Function Approximation agent calls for the Committee Machine agent, which creates the final hybrid models in the form of committees for each of the dependent variables and saves them.

The Regression agent calculates regression models for given independent and dependent variables (see Fig. 4.23). It reads information from belief like data sets for independent variables **X** and dependent variables **Y**. The Regression agent disposes two plans:

- *simpleRegression*, which is used for the generation of univariate regression models, and,
- *multipleRegression*, which is used for multiple regression models generation.

Each plan permits creation of the following linear and non-linear models: hyperbolic, exponential and power models.

The Artificial Neural Network agent creates different types of models based on neural networks (see Fig. 4.23). For this reason, it executes several plans:

- *evaluateImpactAssessment*, which is aimed to calculate environmental impact assessment and to select the most influential factors X for every dependent variable Y, and,
- *neuralNetwork*, which is used for generation of approximation and autoregression models of $Y = F(X)$, $Y = F(t)$ and $X = F(t)$.

Feed-forward neural networks trained with backpropagation algorithm are calculated within the *evaluateImpactAssessment* plan. When an agent chooses the *neuralNetwork* plan, it creates feedforward neural network models trained with:

- Backpropagation algorithm (BP)
- Genetic algorithms (GA)
- Resilient backpropagation algorithm (RPROP)

The Decomposition agent provides a relevant plan *Decomposition* to range the inputs of the neural model by their importance, and illustrates how the model output may change in response to variation of an input (see Fig. 4.23).

The Evaluation agent has four plans for model evaluation. These plans are:

- *evaluateANN*, which is used when the Evaluation agent receives the *StartEvaluation* message from the Artificial Neural Network agent,
- *evaluateSimpleRegression*, which is used when the Evaluation agent receives the *StartEvaluation* message indicating the "univariate" parameter as a regression type from the Regression agent,
- *evaluateMultipleRegression*, which is used when the Evaluation agent receives the *StartEvaluation* message indicating the "multiple" parameter as a regression type from the Regression agent, and,
- *acceptModels*, which is used for each data mining agent in order to check the adequacy of a model from the initial data.

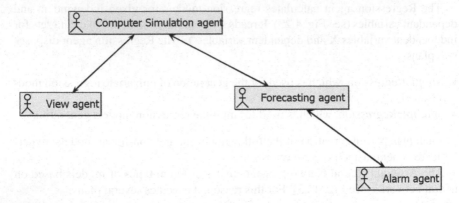

Fig. 4.24 Computer Simulation agent and its team

The Computer Simulation Agent and Its Team

The Computer Simulation agent interacts with the user and performs a set of tasks within the *Computer Simulation*, *Decision Making* and *Data Distribution* roles (see Fig. 4.24). Its subordinate agents are the following ones:

- The Forecasting agent, which is used to create forecasts and predictions of dependent and independent variables.
- The Alarm agent, which is used to identify the values that are received by the Forecasting agent, which exceed permissible level.
- The View agent, which is used to organize the computer-user interaction and create textual, graphical, and other types of documents.

The execution cycle of the Computer Simulation agent begins when it receives a triggering message from the Function Approximation agent (see Fig. 4.25). Next, the Computer Simulation agent asks for the user´s preferences, and, to be more precise, for the information of the disease and pollutants of interest, the period of the forecast, and the ranges of their value changes. Once the information from the user is received, the Computer Simulation agent sends the *SimulateAlternative* message to the Forecasting agent, which reasons and executes one of the plans, which include *Forecasting*, *ModelSimulation*, and *CriterionApplication*. When the alternative is created, the Computer Simulation agent sends the *StartAlarmCheck* message to the Alarm agent. The Alarm agent compares the simulation and forecast data from the Forecasting agent with the permitted and alarm levels for the correspondent indicators. If they exceed the levels, the Alarm agent generates alarm alerts. Fig. 4.26 gives a view of the Computer Simulation agent and its subordinate agents and shows the messages that agents exchange, their plans and the perceptions that they receive.

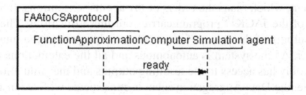

Fig. 4.25 The triggering message from the Function Approximation agent to the Computer Simulation agent

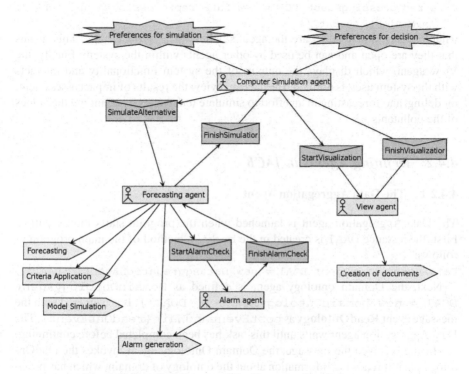

Fig. 4.26 Detailed diagram of the Computer Simulation agent

4.4 Implementation in Jack

4.4.1 Program Architecture

The MAS has an open agent-based architecture, which allows for an easy incorporation of additional modules and tools, enlarging a number of functions of the system. The system belongs to the organizational type, where every agent obtains a class of tools and knows how and when to use them. Actually, such types of systems have a planning agent, which plans the orders of the agents' executions. In our case, the main module of the JACK^{TM}program carries out these functions. The View agent displays the outputs of the system functionality and organizes the interaction with the system user. As the system is autonomous and all the calculations are executed by it, the user only has access to the resulting outputs and the simulation window.

For example, the Data Aggregation agent is constructed with a constructor:

```
DataAggregationAgent DAA1 = new DataAggregationAgent('`DAA'')
```

Its method can be called as *DAA1.fuseData()*. The DataPreprocessingAgent is constructed as follows:

```
DataPreprocessingAgent DCA = new DataPreprocessingAgent('`DPA'',
  `'x.dat'', `'y.dat'')
```

where x.dat and y.dat are the agent's beliefs of a "global" type. This means that they are open and can be used by other agents within the system. Finally, the View agent, which displays the outputs of the system functionality and interacts with the system user, is notified. He/she can review the results of impact assessment, modeling, and forecasting in an effort to simulate tendencies by changing the values of the pollutants.

4.4.2   Defining Agents in JACK

4.4.2.1   The Data Aggregation Agent

The Data Aggregation agent is launched when the program starts the execution. First, the instance DAA1 is created in the body() method of the main program as follows:

```
DataAggregationAgent DAA1 = new DataAggregationAgent('`DAA'').
```

Next, the Domain ontology agent is defined as DomainOntologyAgent DOA1 = new DomainOntologyAgent ('`DOA''). It is triggered with the message event **ReadOntology** as postEventAndWait(readOntEvent).The Data Aggregation agent waits until this task has been completed before continuing.

Having received the message, the Domain Ontology agent invokes the readOntology plan. It reads the information about the ontology of domain, which has previously been created in Protégé. The domain ontology is stored into a *owl*-file named *DomainOntology.owl*. The file contains tags that correspond to the components of the ontology of the domain of interest : Individuals, Classes, Properties, Values, Restrictions and AxiomaticRules (see equation 3.2). Reading classes, their properties, and restrictions from the ontology of domain is supported by the Jena API [99].

First, to read data from the OWL-file, the *readOntology* plan creates a model of the ontology. The complexity of the model is determined by the parameter OntModelSpec.OWL_DL_MEM, which means the chosen language profile is "OWL Description logic (OWL DL)" and the reasoner is not used. OWL DL includes all OWL language constructs with specific restrictions [140]. The ontology model is created as shown in the following code:

```
Model m = FileManager.get().loadModel(''DomainOntology.owl'');
OntModel m2 = ModelFactory.createOntologyModel
   (OntModelSpec.OWL_DL_MEM,m);
```

Then, the plan calls for the readHierarchy(PrintStream out, OntM- odel m) method new DOA.readHierarchy(System.out, m2). This method shows the hierarchy of classes encoded by the given model:

```
public void showHierarchy(PrintStream out, OntModel m){
   Iterator<OntClass> i = m.listHierarchyRootClasses()
      .filterDrop(new Filter<OntClass>(){
            @Override
            public boolean accept(OntClass r) {
                return r.isAnon();
            }});
            while (i.hasNext()) {
            showClass(out,i.next(),new ArrayList<OntClass>(),0);
      }
}
```

The method showClass(PrintStream out, OntClass cls, List <OntClass> occurs, int depth) enables the reading of the OWL-file and creating a list of classes that are contained in the file.

```
protected void showClass(PrintStream out, OntClass cls,
                        List<OntClass> occurs, int depth) {
   renderClassDescription(out,cls,depth);
   out.println();
   if (cls.canAs(OntClass.class)&&!occurs.contains(cls)){
      for (Iterator<OntClass> i=cls.listSubClasses(true);
                        i.hasNext();){
         OntClass sub = i.next();
         occurs.add(cls);
         showClass(out,sub,occurs,depth+1);
         occurs.remove(cls);
      }
   }
}
```

Retrieved class hierarchy is stored by the Domain Ontology agent as a List of OntClass objects.The attributes of classes are also read from the file using the ontology API.

```
DescribeClass dc = new DescribeClass();
if (args.length >= 2) {
   OntClass c = m2.getOntClass( args[1] );
   dc.describeClass( System.out, c);
}
else {
   for (Iterator i= m2.listClasses(); i.hasNext(); )
      dc.describeClass(System.out,(OntClass)i.next());
}
```

When the Domain Ontology agent completes the execution of the *readOntol-ogy* plan, it sends the *OntologyIsBeingRead* message to the Data Aggregation agent. Next, the Data Aggregation agent posts a message event `postEvent (StartFusion)`. This event invokes the following fusion agents: Water Data Fusion agent, the Petroleum Data Fusion agent, the Mining Data Fusion agent, the Traffic Pollution Fusion agent, the Waste Data Fusion agent, and the Morbidity Data Fusion agent. These agents read data from data sources and search for and extract data that corresponds to "Properties" and " Values" components of the ontology of domain. The fusion agents act in a similar way. At the beginning of its execution each fusion agent sends an event **StartReading** in order to choose the appropriate plan. The **StartReading** event is posted with `postEvent()`, the event's posting method, which allows an agent to post an event to itself. The event is handled asynchronously, which enables concurrent execution of plans.

```
public event StartReading extends BDIGoalEvent {
    public String startID;
    #posted as
    startReading(String word)
    {
        startID=word;
    }
    #uses data diseases_region, diseases_region;
    #uses data pollutants_region, pollutants_region;
}
```

In order to choose the appropriate plan, each agent must apply meta-reasoning. With this aim, it determines which plans are handled by the message event reading the `#handles event` statements and then inspects the content of the relevant method of each plan:

```
static boolean relevant(StartReading ev)
    {
        return (ev.startID.equal(word));
    }
```

There are appropriate plans for different data sources:

- data extraction from CSV-files,
- data extraction from DOC and XLS-files, and,
- data extraction from XML-files.

The statement within the `relevant()` method contains the condition to choose plan: the parameter "startID" of the event is equal to the String named "word". The method is posted as:

- `postEvent(startReading(''csv''))` the plan *CSVextraction.plan* is selected,

- `postEvent(startReading(''doc''))` the plan *DOCyXLSextraction. plan* is selected, and,
- `postEvent(startReading(''xml''))` the plan *XMLextraction.plan* is selected.

The fusion agents create data arrays and fill them with extracted data. Then, these data are written into text files. These text files are later converted into beliefs for the Data Preprocessing agent and its subordinate agents. Each data fusion agent finishes its execution by sending the *FusionIsFinished* message to the Data Aggregation agent. Once the agents receive a message from each of the data fusion agents, the Data Aggregation agent terminates its execution. Finally, it generates the triggering message event to the Data Preprocessing agent.

The Data Aggregation agent and its subordinate agents are stored in a nested container "fuse", which has been created in JACK.

4.4.2.2   The Data Preprocessing Agent

The Data Preprocessing agent and its internal architecture, as presented in the Navigator window, are shown in Fig. 4.27. The Data Preprocessing agent is a subclass of the JACK´s Agent class. The "Java" specifications for the agent include the package name, interfaces which it implements, and imported libraries and packages.

First, we need to create an agent's knowledge base. It can be created only during agent´s initialization. That is why the agent´s procedure of reading information from additional plain files and filling in the beliefs is included into agent´s constructor. The Data Preprocessing agent's beliefs include:

1. Beliefs with global access:

   - DISEASES_REGION, which contains data about morbidity,
   - POLLUTANTS_REGION, which contains data about pollutants,
   - DISEASES, which contain descriptions of morbidity classes.

2. Beliefs with private access:

   - POLLUTANTS_REGION_N, which contains normalized data about pollutants,
   - DISEASES_REGION_N, which contains normalized data about morbidity.

Beliefs with global access can be shared among all agents in the process. They can also have private access represented by data of the agent instance. The Data Preprocessing agent has permissions to read, write, change and delete tuples from the beliefs DISEASES_REGION, POLLUTANTS_REGION, and DISEASES, and shares them with the Data Aggregation agent. Beliefs with private access of the Data Preprocessing agent are used for data normalization.

Second, the reasoning elements of the agent are written into the "Cleaning" capability. It includes:

- triggering event **StartDataConsistencyCheck**, which the Data Preprocessing agent sends to the GapsAndArtifactsCheck agent;

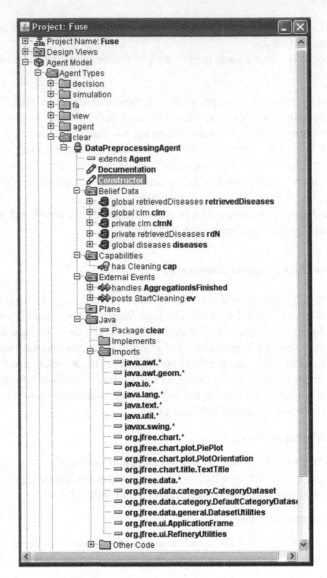

Fig. 4.27 The Data Preprocessing agent and the view of its internal structure. JACK Navigator View

- triggering event **StartSmoothing**, which the Data Preprocessing agent sends to the DataSmoothing agent;
- triggering event **StartNormalization**, which the Data Preprocessing agent sends to the Normalization agent;
- triggering event **StartCorrelation**, which the Data Preprocessing agent sends to the Correlation agent;

- beliefsets DISEASES_REGION, POLLUTANTS_REGION, DISEASES,POLLUTANTS _REGION_N, and DISEASES_REGION_N.

The GapsAndArtifactsCheck agent executes the *EliminateArtifacts* plan. The *EliminateArtifacts* plan, resolves two tasks: it searches for outliers and eliminates them, then fills the missing values. First, the GapsAndArtifactsCheck agent reads data from the POLLUTANTS_REGION and the DISEASES_REGION beliefs and creates data arrays. The method of the interquartile range (*IRQ*) is used to identify outliers. With this aim, the values of the 25-th, the 75-th percentile and the $IQR = Q3 - Q1$, are calculated for each variable. Then, each value from the data set is compared with boundary values, and if the value lies beyond the interval $[(Q1 - 1.5(IQR), (Q3 + 1.5(IQR)]$, its value is deleted from the correspondent data array (see subsection 3.6.1.2). As a result, the artifacts are eliminated on this step of the plan execution. Next, the missing values within the data set data set for each variable (pollutant and disease) are filled with "the golden ratio" of the neighboring values. The "golden ratio" is calculated with the step 3 as $(0.38 \cdot (value \, previous \, to \, the \, gap) + 0.62 \cdot (value \, posterior \, to \, the \, gap))/2$ (see Section 3.6.1.2). If the missing value is the first or the last value in the data set, it is filled with the value of the posterior or previous value respectively. The new value is written to the corresponding data array.

After that, the *EliminateArtifacts* plan completes the execution and the GapsAndArtifactsCheck agent updates the POLLUTANTS_REGION and the DISEASES _REGION beliefs, and sends the *ConsistencyIsBeingChecked* message to the Data Preprocessing agent.

The DataSmoothing agent is triggered by the *StartSmoothing* message from the Data Preprocessing agent. As soon as the message is received, the DataSmoothing agent initializes, and then initiates the *Smooth* plan, which uses exponential smoothing with α equal to 0.15. The agent has access to read and to modify the POLLUTANTS_REGION and the DISEASES_REGION beliefs. It creates data arrays and organizes cycles for each variable. Within the cycle, each data value is smoothed in agreement with equation (3.9). Next, the DataSmoothing agent modifies the POLLUTANTS_REGION and the DISEASES_REGION beliefs.

The Normalization agent also has access to modify the POLLUTANTS_REGION and the DISEASES_REGION beliefs. When the Normalization agent is triggered by the *StartNormalization* message from the Data Preprocessing agent, it invokes the *Normalize* plan. As with the GapsAndArtifactsCheck and the DataSmoothing agent, the Normalization agent creates arrays with data and creates a cycle for each variable. Within the cycle, it calculates two types of normalization: the Z-score standardization and the "Min-Max" normalization (see equations 3.10 and 3.11). The Normalization agent modifies the POLLUTANTS_REGION and the DISEASES_REGION beliefs with data normalized with the Z-score standardization method. The data normalized with the "Min-Max" normalization method are written into private beliefs POLLUTANTS_REGION_N, and DISEASES_REGION_N. These beliefs are used by the Artificial Neural Network agent. Having finished creating and updating of the beliefs, the Normalization agent sends the *NormalizationIsFinished* message.

The message event **StartCorrelation** from the Data Preprocessing agent invokes the Correlation agent. It reads the POLLUTANTS_REGION and the DIS-EASES_REGION beliefs and creates two arrays: "X" and "Y". The Correlation agent has two plans: "Parametric Correlation" and "Non-Parametric Correlation". When the message comes, the agent has to choose the plan, as both plans are applicable. Since the message event is a BDIMessageEvent, the agent has the option of performing meta-level reasoning to determine which plan is best. For this reason, the Correlation agent has the syntactic relevance:

- identifying the plans which handle the event type checking the #*handles event . . .* statement,
- inspecting data which is contained in the relevant() method to get additional information regarding the event,
- defining the set of all applicable plans checking the context() method to analyze information stored as part of the agent's beliefs.

The correlation matrices are calculated by the Correlation agent and stored as ParamCorrelation.dat and NonParamCorrelation.dat files. These files are read by the Function Approximation agent during its initialization and are converted to its beliefs.

The Data Preprocessing agent and its subordinate agents are guarded in a nested container "clear" that has been created in JACK.

4.4.2.3 The Function Approximation Agent

The Function Approximation agent is responsible for data mining. Its general structure is given in Fig. 4.28. After initialization, it posts the StartMining event to read data from beliefs DATAX, DATAY and FORMODELS:

```
#reads data dataX dataXfa;
#reads data dataX dataYfa;
#reads data forModels forModelsFA;
```

In the following part of the code, the Function Approximation agent fills the following arrays with data from its beliefs:

```
double [][][] dataXfapp =
 new double [region.length][pollutants.length][years.length];
double [][][] dataYfapp =
 new double [region.length][diseases.length][years.length];
int forModelsFapp[][][] =
 new int [region.length][diseases.length][models_number];
```

To begin reading data from beliefs, the Function Approximation agent uses the createDataArrays() method:

```
public void createDataArrays()
    {
      postEventAndWait(ev.start(''''));
    }
```

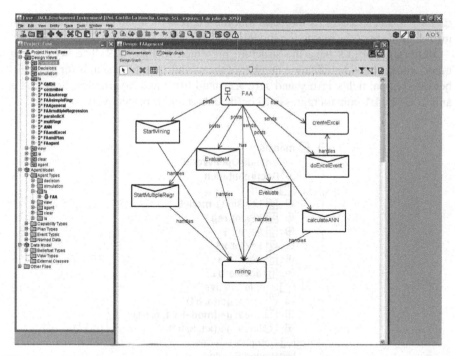

Fig. 4.28 The general view of the Function Approximation agent, its capabilities and its events

The **StartMining** event handles the *createData* plan. Since the **StartMining** event was sent using the `postEventAndWait` command, the Function Approximation agent waits until the relevant plan finishes its execution. Then, the Function Approximation agent sends the *StartDataMining* message to subordinate agents, which initiate execution of the relevant data mining plans.

When the **Regression agent** receives the message from the Function Approximation agent, it sends message events to invoke relevant plans.

```
#sends event StartMining  ev1;
```

The relevant plan for univariate linear regression calculation is *simpleRegr*. It declares usage of named data:

```
#handles event StartMining ev;
#uses interface Function Approximation agent self;
#uses data dataX dataXfa;
#uses data dataX dataYfa;
#uses data forModels forModelsFA;
#modifies data models models;
```

Here DATAXFA, DATAXFA and FORMODELSFA are global beliefs. RA has permission to read information from these beliefs. The last statement #modifies `data models models` indicates that the Regression agent creates beliefs with information about the regression model.

Beliefset MODELS have the structure presented in Fig. 4.29. The field String model contains an identification number of the model; `int reg` stands for the number of a region; fields `int y` and `int x1` stand for Y and X variables; `double a` and `double b1` contain regression coefficients a and b respectively.

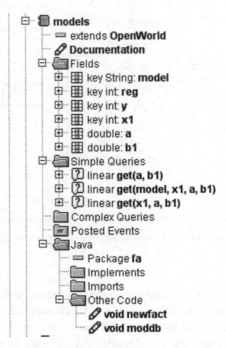

Fig. 4.29 The internal structure of beliefset MODELS

There are three simple linear queries for data stored in this beliefset;

1. getting regression coefficients a and b,
2. getting a number of the `model`, a number of variable X and regression coefficients a and b, and,
3. getting number of variable X and regression coefficients a and b.

When the Regression agent requests relevant plans, it sends the **Evaluate** event:

```
postEventAndWait(ev1.evaluate(slovo))
```

where `slovo` is a String variable, which stands for a type of regression model. Depending on value of `slovo`, one of the following plans can be chosen:

- *simpleRegr*, or,
- *multipleRegr*.

The plan **simpleRegr** calculates coefficients for univariate regression and uses the method:

```
uniRegression(double [] X, double [] Y, String slovo),
```

where double arrays X and Y contain data from DATAXFA and DATAYFA named data sets. The argument String slovo indicates the type of regression: linear, hyperbolic, power or exponential. The method uniRegression is based on the calculation approach, given in Chapter 3.6. After regression models are created, their characteristics are added to the MODELS beliefset:

```
if (ev.type.equals(''simple''))
  models.add(''simple'',i,j,factor,self.A[i][j][k],self.B[i][j][k]);
if (ev.type.equals(''power''))
  models.add(''power'',i,j,factor,self.A[i][j][k],self.B[i][j][k]);
if (ev.type.equals(''expon''))
  models.add(''expon'',i,j,factor,self.A[i][j][k],self.B[i][j][k]);
if (ev.type.equals(''hyperb''))
  models.add(''hyperb'',i,j,factor,self.A[i][j][k],self.B[i][j][k]);
```

The plan *multipleRegr* is identical to the plan *simpleRegr*. It is invoked by the **StartMining** event in concurrent mode with *simpleRegr* plan. The plan *multipleRegr* handles the event **StartMultipleRegr ev**, and uses the following beliefsets:

```
#uses data dataX dataXfa;
#uses data dataX dataYfa;
#uses data forModels forModelsFA;
#modifies data modelsM modelsM;
```

During the plan execution, the belief MODELSM is created. It has the structure, shown in Fig. 4.30.

The structure of the MODELM beliefset is the same as the structure for the beliefset MODELS, though fields x2 and b2 (which stand respectively for X2 variable and its regression coefficient) are added. There are three simple linear queries to this beliefset, which allow get the necessary parameters of any multiple regression model created. The Function Approximation agent uses the method:

```
multiRegression(double [] Y,double [] X1, double [] X2,
    String slovo)
```

to calculate multiple regression models. When the Regression agent completes execution of regression models, it sends the *StartEvaluation* message to the Evaluation agent. Having received the *StartEvaluation* message from the Regression agent, the Evaluation agent invokes relevant plans:

- *evaluateSimple* for univariate regression models;
- *evaluateMultiple* for multiple regression models evaluation.

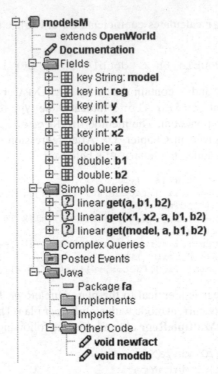

Fig. 4.30 The internal structure of MODELSM beliefset. View in the JACK environment

The plans *evaluateSimple* and *evaluateMultiple* read information about models from beliefsets MODELS and MODELM, evaluate the models and, finally, fill in datasets with the accepted models. For this reason, the Evaluation agent creates two beliefsets: MODELSCRITERIA ans MODELSMCRITERIA. Fig. 4.31 presents the JACK navigator view for both datasets.

The fields of the MODELSCRITERIA beliefset contain the following information: `model` is an identification of regression model type ("linear", "hyperb", etc.) from the `models` named data; int `reg` stands for a number of a region; fields int `y` and int `x1` stand for Y and X variables; double `a` and double `b1` contain regression coefficients a and b respectively, double fields `correl`, `determ`, F stand for correlation and determination coefficients and for *F*-value for the Fisher test.

The MODELSMCRITERIA beliefset has the same fields as the `modelsCriteria` beliefset with the exception that `model` is an identification of multivariate regression model type ("linear", "giperb", etc.) from the MODELSM named data. The fields `x2` and `b2` (which stand respectively for X2 variable and the correspondent regression coefficient) are added.

The **Artificial Neural Network agent** executes various plans to create models based on artificial neural networks. With this aim the Encog library [42] is used. The Encog library contains classes which allow to create a wide variety of networks (Feedforward Neural Network, Hopfield Neural Network, Radial Basis Function

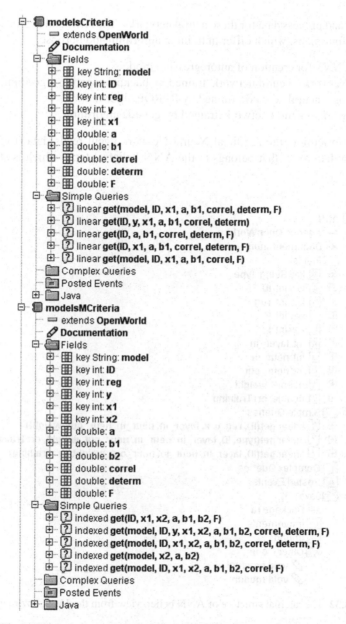

Fig. 4.31 The JACK navigator view of beliefsets MODELSCRITERIA and MODELSMCRITE-RIA

Network, etc.) and to train them with various algorithms (Backpropagation, Resilent propagation, Genetic algorithms, etc.). The library also includes support classes to normalize and process data for these neural networks. The Artificial Neural Network agent has four plans, which differ in training algorithm and purpose:

- *autoregNN* - for creation of autoregression models,
- *neuralNetwork* - neural network, trained by backpropagation algorithm;
- *RNencog* - neural network, trained by RPROP algorithm,
- *GAnetworks* - neural network, trained by genetic algorithms.

During execution, the Artificial Neural Network agent reads data from beliefs and fills in data ANN, that belongs to the ANN beliefset type, which is shown on Fig. 4.32.

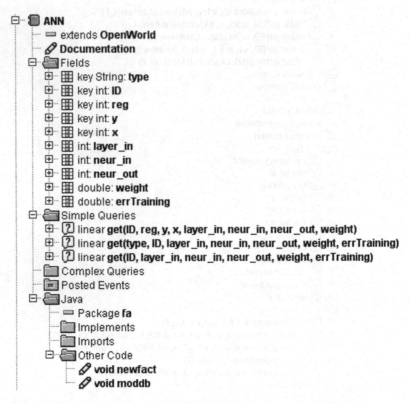

Fig. 4.32 The internal structure of ANN belief, view from the JACK environment

Beliefs of the ANN type have the following structure: the field of the String type contains a type of the neural network model (BP stands for a neural network trained with the backpropagation algorithm, RPROP stands for a neural network trained with the resilient propagation algorithm, GANN stands for a neural network trained with genetic algorithms); int ID stands for the identification number of a model (is a key field); int reg stands for the number of a region; fields int y and int x stand for Y and X variables; the fields int *layer_{in}*, *neur_{in}* and *neur_{out}* contain the number of the previous layer of ANN, the number of a neuron in this layer and the number of a neuron in the next layer; and double weight and errTraining contain weight and training error, respectively. Plans are triggered by *calculateANN* event. Outcomes are written into ANN belief.

When neural network models are calculated, the Artificial Neural Network agent sends the *Start Evaluation* message. This message triggers the Evaluation agent, that launches the *evaluateANNs* plan. When the Evaluation agent completes this plan, it generates the *EvaluationIsFinished* message for the Artificial Neural Network agent.

The Committee Machine agent only has one plan *createCM*. This agent reads information about all created and accepted models from datasets ANN (which contains error of training for each ANN), MODELS2 (which contains approximation criteria for univariate regressions) and MODELSM2 (which contains approximation criteria for multiple regressions). The Committee Machine agent searches for and retrieves models with the best criteria values for each dependent variable **Y**, and combines them into one hybrid model. As a result, the Committee Machine agent creates the dataset COMMITTEE, which has the following fields:

```
public beliefset commitee {
    #key field int Ncommitee;
    #key field int reg;
    #key field int DisOrPollut;
    #key field int NofDisOrPollut;
    #value field int model1;
    #value field int model2;
        . . .
    #value field int modelN;
}
```

Here, a beliefset contains the identification numbers of models, which are included into a final hybrid model.

The following code shows how the Committee Machine agent searches for models for the variable *j*. It scans beliefs MODELS2 and MODELSM2, using "get" queries. If the model is found, its determination coefficient is added to the array list selected. Rvk is the coefficient of determination for the selected model.

```
//look through all the tuples in models2 beliefset
//and select models for given variables i,j and fact
for (Cursor c = models2 get(model,ID,i,j,fact,a,b1,
        correl,determ,F); c.next(); ) {
        Rkv=determ.as_double();
        selected.add(new Double(Rkv));
    }
for (Cursor c = modelsM2.get(model,ID,i,j,fact1,fact2,
        a,b1,b2,correl,determ,F); c.next(); ) {
```

```
            Rkv=determ.as_double();
            selected.add(new Double(Rkv));
    }
```

Next, records of array list are sorted and the best models are selected for a hybrid model.

4.4.2.4   The Computer Simulation Agent

The Computer Simulation agent interacts with the user and receives his/her preferences for forecasting and simulation. It also manipulates the Forecasting agent, the Alarm agent and the View agent. The Computer Simulation agent is created with the command:

```
ComputerSimulationAgent DCA = new
ComputerSimulationAgent (''CSA'', ``x.dat'', ``y.dat'',
    ''selectedModels.dat'', ``criticalValues.dat'')
```

where CSA is the name of the agent, dataX.dat and dataY.dat are the files that contain initial data, selectedModels.dat contains information about selected models and criticalValues.dat contains information about critical levels for factors. These text files are used in the Computer Simulation agent beliefs creation during its initialization.

The Computer Simulation agent sends a message *SimulateAlternative* and triggers the Forecasting agent, which, in its turn, generates the message *startSimulation* with the String parameter start. This message invokes plan *simulate*:

```
public plan simulate extends Plan {
    #handles event startSimulation ev;
    #uses data modelsMCriteria modMultiple;
    #uses data modelsCriteria modSimple;
    #uses data ANN ANN;
    #uses data committee comMach;
```

```
static boolean relevant(startSimulation ev)
{
    return (ev.start.equals(''start''));
}
```

The plan *simulate* calculates forecasting values for the chosen variable of interest. First, it reads from the COMMACH beliefs information about models, which are included into the committee machine hybrid model for the chosen variable of interest **Y**. Second, the Forecasting agent searches for these models in corresponding beliefs: MODSIMPLE, MODMULTIPLE and ANN. Finally, the Forecasting agent approximates output values for the hybrid model.

The Alarm agent contains data about pollutants and their permissible levels. After approximation and forecasting, the Forecasting agent generates the *StartAlarm-Check* message and sends it to the Alarm agent. If forecasted values overcome permissible levels, the Alarm agent generates an alarm message, which is delivered to the View agent.

4.4.2.5 The View Agent

The View agent is implemented by the command:

```
ViewAgent VA = new ViewAgent(''VA'',''regions.dat'',
    ''years.dat'',''pollutants.dat'',''diseasesEn.dat'',
    ''forModels.dat'',''ages.dat'',''xNew.dat'',
    ''yNew.dat'',''yPred.dat'',''ranks.dat'');
```

where VA is a name of the agent; regions.dat is the textual file, which contains names of regions; years.dat is the textual file, that contains numbers of years; pollutants.dat is the textual file, which contains names of pollutants; diseasesEn.dat is the textual file, which contains names of diseases; forModels.dat contains links to the best models that were created, ages.dat is the textual file, which contains information about ages; xNew.dat contains X data sets with pollutants; yNew.dat contains Y data sets with diseases; yPred.dat contains approximated Y data sets; ranks.dat contains the results of the correlation matrix for X and Y. All this knowledge from the View agent´s constructor is transformed into its beliefs.

The View agent contains both numerical and textual data with the aim to organize the human - computer interface and provide the user with all the necessary information.

Fig. 4.33 gives a view of the principal windows of the View agent. The window marked "1" is a principal window, which shows a work flow carried out by agents. These works correspond to the logical levels of the agent-based decision support system and include:

- Data retrieval and fusion
- Data clearing
- Modeling
- View results
- Simulation

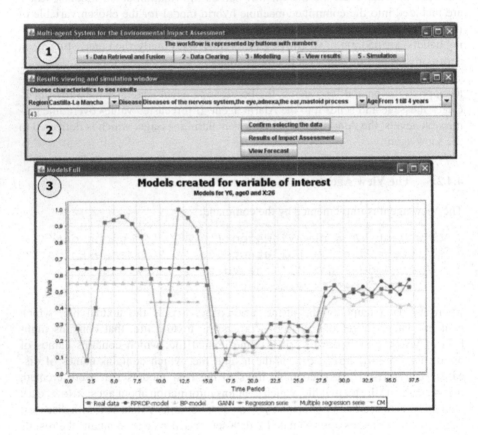

Fig. 4.33 The cascade of windows realized by the View agent

The second window gives a view of the simulation window and in it marked with a "2". There is the option to choose a region, a disease type, and an age range, and shows results of impact assessment and forecasting. The last view provided (marked as "3") shows the committee machine model for selected disease "Diseases of the nervous system, eye, adnexa, the ear, mastoid process", for the selected region "Castilla-La Mancha", and for the specific age group "from 1 to 4 years". The chart shows models which have been included in the committee, and are defined in the legend field.

Chapter 5
Data and Results

Creativity is a power to connect the seemingly unconnected.
William Plomer

Abstract. The case study introduced in the previous chapter was performed in order to apply the DeciMaS framework for the identification and evaluation of environmental impact upon human health and generation of alternative decisions sets. In this chapter, a computational experiment is carried out by means of an agent-based decision support system, which sequentially executes and completes each stage of the DeciMaS framework. Thus, this chapter is dedicated to the discussion of the obtained results of the experiment for the case of environmental impact for the selected regions. Data and experiment results of data modeling, simulation, sensitivity analysis, impact assessment and decision generation are discussed.

5.1 Introduction

The experiment designed to evaluate the possible harm caused by environmental contamination upon public health has been conducted in the region of Castilla-La Mancha. The increasing public health threats tend to affect the physical, chemical, and biological nature of our natural environments as well. From a technical point of view, the experiment aims to demonstrate the possibilities of the DeciMaS framework to model complex systems and to create decision support systems, which are able to assist a decision maker.

The case of the human health impact assessment appeared to be an optimal domain for application. First, the information is scarce and heterogeneous. The heterogeneity of the initial data is multiple as various parameters (for example, morbidity indicators versus waste subgroups) have different amounts of data available. The periods between registration of parameters are different. For example, one parameter is registered on a monthly, and the other on a yearly basis. Some data are measured in different scales, for example, morbidity is measured in persons and in thousands of persons. Second, data sets appear to be short, and it has been decided to apply special data mining methods, such as GMDH-based models, and committee machines. Last but not least, the domain of the study represents one of the most difficult problem areas. The real interrelations between its components have not been thoroughly

M.V. Sokolova, A. Fernández-Caballero: Decision Making in Complex Systems, ISRL 30, pp. 139–167.
springerlink.com

studied yet, even by domain experts. That is why the application of the DeciMaS methodology may have very effective results as it facilitates the discovery of new knowledge as well as a new understanding of the nature of complex systems.

5.2   Data for Experiment

To evaluate the impact of environmental parameters upon human health in Castilla-La Mancha, retrospective data dated from 1989 until 2007, has been used. Resources offered by the Instituto Nacional de Estadística and by the Instituto de Estadística de Castilla-La Mancha are used for the research [87]. The factors, which describe the "Environmental pollution - Human health" system, are used as indicators of human health and the influencing factors of environment. These can cause negative effects upon the indicators of human health, noted above. The factors used in the experiment are presented in Table 5.1.

Morbidity, classified by sex and age, is accepted as an indicator to evaluate human health. Table 5.2 gives a list of diseases examined in this case study. The diseases included in the research are chosen in accordance with the International Statistical Classification of Diseases and Related Health Problems (ICD) [83]. The sex groups included the following ones: "males", "females" and "total"; and the age groups consist of:

- all the ages;
- under 1 year;
- 1 - 4 years;
- 5 - 14 years;
- 15 - 24 years;
- 25 - 34 years;
- 35 - 44 years;
- 45 - 54 years;
- 55 - 64 years;
- 65 - 74 years;
- 75 - 84 years;
- 85 years and over.

Information is retrieved from CSV, DOC and XLS-files and fused together. An example of an XLS-file, which contains information about wastes, generated in 2002, is provided in Fig. 5.1. Wastes are classified by territory and divided into smaller groups, depending on contamination hazard, origin and provenance. Four columns have numerical values for dangerous and non-dangerous wastes for the region of Castilla-La Mancha and for Spain. The unit of measure, "tonnes", is also given in the file. Values of morbidity are retrieved from the XLS-files.

An example of such a file is also presented in Fig. 5.1. In this case the source file is an MS Excel book that consists of several pages: *1993, 1994, 1995, 1996, 1997, 1998* which are related to correspondent years (1993-1998 years). Here, each page contains information of interest: classes of diseases and mortality value for each class and each age group. In this case, the agent must revise each page to retrieve data.

Table 5.1 Pollutants studied in research

Type of Disease / Pollutant	Disease class
1 Transport	Number of Lorries, Buses, Autos, Tractors, Motorcycles, Others
2 Usage of petroleum products	Petroleum liquid gases; Petroleum autos; Petroleum; Kerosene; Gasohol; Fuel-oil
3 Water characteristics	Chemical oxygen demand (COD); Biochemical oxygen demand (BOD$_5$); Solids in suspension; Nitrites
4 Wastes	Non-dangerous chemical wastes; Other non-dangerous chemical wastes; Non-dangerous metal wastes; Wastes from used equipment of paper industry; Dangerous wastes of glass; Dangerous wastes of rubber; Dangerous solid wastes; Dangerous vitrified wastes; Wastes from used equipment; Metallic and phosphorus wastes
5 Principal miner products	Hulla/Hull; Mercury; Kaolin; Salt; Thenardite; Diatomite; Gypsum; Rock; Others

Table 5.2 Diseases studied in research

Type of Disease / Pollutant	Disease class
1 Endogenous diseases	Certain conditions originating in the perinatal period Congenital malformations, deformations and chromosomal abnormalities
2 Exogenous diseases	Certain infectious and parasitic diseases Neoplasm, Diseases of the blood and blood- forming organs and certain disorders involving the immune mechanism Endocrine, nutritional and metabolic diseases Mental and behavioral disorders, Diseases of the nervous system Diseases of the eye and adnexa, Diseases of the ear and mastoid process Diseases of the circulatory system, Diseases of the respiratory system Diseases of the digestive system, Diseases of the skin and subcutaneous tissue Diseases of the musculoskeletal system and connective tissue Diseases of the genitourinary system, Pregnancy, childbirth and the puerperium Symptoms, signs and abnormal clinical and laboratory findings, not elsewhere classified External causes of morbidity and mortality

1.3.13. RESIDUOS GENERADOS POR ACTIVIDAD ECONÓMICA Y TIPO DE RESIDUO.
INDUSTRIAS EXTRACTIVAS, MANUFACTUREREAS, ENERGÉTICAS Y CONSTRUCCIÓN.

Toneladas

	Castilla-La Mancha		España	
Año 2002	No peligrosos	Peligrosos	No peligrosos	Peligrosos
Total residuos	645.137	16.164	57.464.734	1.575.538
01. Residuos químicos (no incluye 01.3)	418	4.363	1.630.248	491.751
01.3 Aceites usados	871	1.026	17.328	88.818
02. Residuos de preparados químicos	163	1.896	156.999	292.783
03. Otros residuos químicos	22.773	2.678	1.027.210	243.976
05. Residuos sanitarios y biológicos	67	30	12.798	2.525
06. Residuos metálicos	11.340	493	2.464.003	113.275
07.1 Residuos de vidrio	6.740	0	291.250	0
07.2 Residuos de papel y cartón	18.463	0	1.071.900	0
07.3 Residuos de caucho	162	0	56.662	0
07.4 Residuos de plástico	2.747	..	198.135	..
07.5 Residuos de madera	49.478	15	507.289	334
07.6 Residuos textiles	620	492	101.121	6.830
08. Equipos desechados	253	4.734	27.982	33.312
09. Residuos animales y vegetales	112.581	0	1.580.339	0
10. Residuos corrientes mezclados	11.291	0	1.711.715	0
11. Lodos comunes	19.747	0	3.015.109	0
12. Residuos minerales y de la construcción (no incluye 12.4)	387.297	439	40.303.369	53.191
12.4 Residuos de la combustión	69	0	3.029.682	248.745
13. Residuos solidificados y vitrificados	58	0	261.595	0

Fuente: Encuesta sobre generación de residuos en el sector industrial. INE.

	A	B	C	D	E	F	G	H	I	J	K	L	M
2													
3		Todas las edades	Menos de 1 año	De 1 a 4 años	De 5 a 14 años	De 15 a 24 años	De 25 a 34 años	De 35 a 44 años	De 45 a 54 años	De 55 a 64 años	De 65 a 74 años	De 75 a 84 años	De 85 y más años
4	Total												
5	I-XVII TODAS LAS CAUSAS												
6	Total	16.531	99	39	44	172	241	282	502	1.388	3.453	5.528	4.783
7	I.01-18 ENF. INFEC. Y PARAS.												
8	Total	216	3	7	2	7	5	2	5	15	45	76	49
9	01 COLERA												
10	Total	..	..	..	..	..	..	..	..	..	..	..	..
11	02 FIEBRE TIFOIDEA												
12	Total	1	..	..	..	..	..	..	..	1	..	..	..
13	03 OTRAS INFEC. INTESTINAL.												
14	Total	15	..	..	..	..	..	..	1	..	2	4	8
15	04-05 TUBERCULOSIS												
16	Total	23	..	..	..	..	1	..	1	2	6	11	2
17	04 DEL AP. RESPIRATORIO												
18	Total	17	..	..	..	..	1	..	..	2	4	9	1
19	05 OTRAS TUBERCULOSIS												
20	Total	6	..	..	..	..	..	..	1	..	2	2	1
21	06 TOS FERINA												
22	Total	..	..	..	..	..	..	..	..	..	..	..	..
23	07 INFEC. MENINGOCOCICAS												
24	Total	4	1	2	1	..	..	..	..	..	..	..	..

1993 / 1994 / 1995 / 1996 / 1997 / 1998

Listo

Fig. 5.1 Examples of an *.*xls* file with information about generated wastes and *.*xls* book with information about morbidity (original files in Spanish)

5.2.1 Data Retrieval and Fusion

There are 148 data files that contain information of interest. These files are used in order to extract the information. After extraction, data is placed in data arrays in accordance with ontology. The fusing agents, which are supervised by the Data

extractCSV.plan

> Input: Data Sources *.csv.
> Output: Retrieved information **C** and **P**.

> (1) Read file line to line and analyze tokens. Search for elements **C** and **P**. Extract found concepts/properties and "remember" their positions in the file.
> (2) Search for concepts/properties as intersections of found concepts.
> (3) Change found properties if their scales are different from the standard ones.
> (4) Write retrieved concepts **C** and **P** to internal file/DB and send it to **DAA**.

extractDOC_XLC.plan

> Input: Data Sources *.doc, *.xls.
> Output: Retrieved information **C** and **P**.

> (1) Search for elements from **C** and **P** in rows or/and columns. Mark found concepts/properties.
> (2) Search for concepts/properties as intersections of found rows and columns.
> (3) Change found properties if their scales are different from the standard ones.
> (4) Send retrieved concepts **C** and **P** to **DAA**.

Fig. 5.2 The plans for data extraction from the CSV, XLS and DOC files

Aggregation agent, works with CSV, XLS and DOC-files. The algorithm of data extraction from these types of files is given in Fig. 5.2.

The Domain Ontology agent has read the OWL-file that contains ontology of the system. First, the agent creates the hierarchy of classes:

```
Class :Pollution
    Class :Water_pollution
    Class :Solar_radiation
    Class :Transport
    Class :Dangerous_wastes
        Class :Urban_waste_products
        Class :Industrial_waste_products
    Class :Industry
        Class :Minery_products
Class :individuals_pollution
Class :Morbidity
    Class :Exogeneous
    Class :Endogeneous
Class :Ontology_of_Environment
Class :Data
Class :Region
```

The part of the output, given above, shows the names of the classes from the OWL-file. Next, the hierarchical links between the classes, shown with "sub-class" and "super-class" properties are also retrieved. For example the line Pollution is a super-class of Class :Water_pollution shows one such relation. The restrictions and properties of each class are shown in this part of the output, generated by the Domain Ontology agent:

```
Class:Ontology_of_Interactions
        is a sub-class of owl.Thing
    Restriction with ID a-5
    on property :has_initiator
    some values from Class :Ontology_of_Agents
  is a sub-class of Restriction with ID a-6
    on property :has_receiver
    some values from Class :Ontology_of_Agents
  is a sub-class of Class owl:Thing
Class :Ontology_of_Agents    is a sub-class of is
        a sub-class of owl.Thing
    on property :has_believes
    some values from Class :Data
  is a sub-class of Restriction with ID a-8
    on property :has_desires
    some values from Class :Methods
  is a sub-class of Restriction with ID a-9
    on property :has_intentions
    some values from Class :Methods
  is a sub-class of Class owl:Thing
Class :Ontology_of_TASKS  is a sub-class of is a
            sub-class of owl.Thing
    on property :has_method
    some values from Class :Methods
  is a sub-class of Class owl:Thing
  is a super-class of Class :Methods
```

The ontology data is converted into agents' beliefs, which is used for further data retrieval.

5.3 Information Fusion and Preprocessing

5.3.1 Detection and Elimination of Artifacts

Data is checked for the presence of missing values and outliers. These can be caused by registration errors or misprints. First, data sets are checked for the presence of

Table 5.3 The results of the missing values and outliers detection

Factor	Missing values, %	Outliers %	Factor	Missing values, %	Outliers %
X_0	38	0	X_1	78	0
X_2	38	0	X_3	40	0
X_4	87	0	X_5	87	0
X_6	60	0	X_7	0	0
X_8	0	0	X_9	0	0
X_{10}	0	0	X_{11}	0	0
X_{12}	30	0	X_{13}	0	6
X_{14}	0	6	X_{15}	0	0
...	...	...	...	...	...
X_{44}	18	0	X_{45}	12	0
X_{46}	0	12	X_{47}	0	12
X_{48}	6	6	X_{49}	18	0
X_{50}	18	0	X_{51}	0	0
X_{52}	12	0	X_{53}	0	0
X_{54}	0	0	X_{55}	37	0
X_{56}	25	0	X_{57}	18	0
X_{58}	18	0	X_{59}	25	0
X_{60}	25	0	X_{61}	25	0
...	...	...	...	...	...
Y_0	12	0	Y_1	12	0
Y_2	0	6	Y_3	0	6
Y_4	8	0	Y_5	0	0
...	...	...	...	...	...

missing values. It is discovered that 13 out of 65 factors have more than 50% of gaps, because these factors had not been registered until several years ago. For example, data for the pollutants "Solids in suspension" and "Nitrites" has been available since 1996, and data for some types of wastes such as "Dangerous wastes of paper and carton" and "Dangerous chemical wastes" do not have records from 1989 until 1998. As a result, the number of pollutants valid for further processing decreases from 65 to 52. Moreover, inconsistent data sets are excluded from the analysis.

Next, pollution indicators are checked for the presence of the outliers and the results are provided in Table 5.3. The human health indicators appear to be more homogeneous, and there are more data sources that contain information of interest. These data do not contain missing values.

5.3.2 Filling of Missing Values

As artifacts are eliminated in the previous step, they are marked as missing values or gaps. The presence of missing values skews the data and may lead to incorrect or unreliable conclusions. In the current study, some data sets have a lot of gaps. The

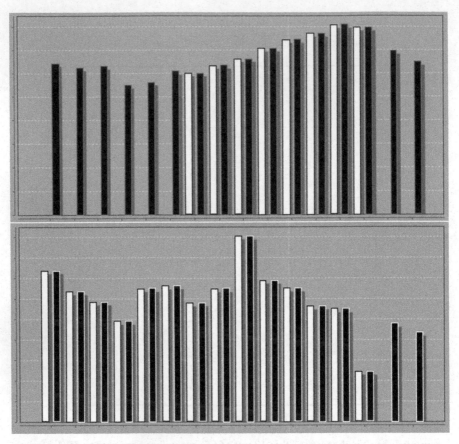

Fig. 5.3 The bar chart that exemplifies the filling gaps procedure: the data before (in white), and the data after filling the gaps (in black)

bar chart given in Fig. 5.3 visualize the filling gaps procedure for given data sets before (in white) and after (in black).

5.3.3 Smoothing Results

The reason to apply smoothing is to homogenize data after the treatment of missing values. The exponential smoothing with the coefficient α equal to 0.15 is used. This value of α provides a "light" smoothing.

5.3.4   Normalization Results

Data is normalized using two normalization methods: Z-score standardization and the "Min-Max" normalization. The example from Fig. 5.4 shows the output of the normalization. The column "Normalized X" contains values normalized within the interval [0, 1]. The extremal values of the real data are shown below as the minimum and the maximum value of the data set.

5.3.5   Results of the Correlation Analysis

Data sets fused from data sources are short and each one contains 32 values. Owing to this fact, non-parametric correlation coefficients are calculated using Kendall´s τ statistic. For the given data sets the critical value of Kendall´s τ is equal to 0.6226. The correlation analysis has obtained the following outcomes. In the same data pool, the variables correlating significantly with morbidity from "Neoplasm" for the age group "under 1 year" are "Water characteristics" and "Wastes from used equipment"; for the age group "more that 85 years", apart from the same factors, are "Wastes: non-dangerous and dangerous chemical waste" (Kendall´s τ-0.726 and -0.983); for the age group "more than 1 and less than 4", significant correlation is found with "'Usage of petroleum products: gases" (Kendall´s τ-0.650), with "Wastes" (Kendall´s τ -0.850 - 1.0).

In relation to endogenous diseases, the outcomes of the "Certain conditions originating in the perinatal period" correlation show relations with "Water characteristics" (Kendall´s τ-0.650), "Principal miner products"(Kendall´s τ-0.750), "Non-dangerous waste from chemical substances"(Kendall´s τ-0.733), and "Metallic and phosphorus wastes"(Kendall´s τ-0.750). The data from the class, "Congenital malformations, deformations and chromosomal abnormalities" correlates with "Principal miner products"(Kendall´sτ-1.0), "Dangerous wastes of paper, glass and rubber"(Kendall´s τ-1.0), and with "Dangerous solid and vitrified wastes"(Kendall´s τ-0.767). The closer the Kendall´s τ value to [1], the stronger the agreement between the two rankings of the analyzed variables. The closer the Kendall´s τ value to [-1], the stronger is the disagreement between the two rankings of the analyzed variables. For these reasons, the calculated values of non-parametric correlation for the above mentioned variables proves the existence of statistical similarities between them, as their the Kendall´s τ coefficients´ values are greater than the critical value 0.6226.

During the testing, an interface window is created, which contains the final results of data cleaning. Fig. 5.4 offers the changes in data after the detection of outliers, filling gaps, and normalization (above). The bar chart provided in Fig. 5.3 shows the data set before and after filling the gaps.

In conclusion, knowledge about the domain of interest has been successfully retrieved from the OWL-file. Next, data has been successfully retrieved from various containers, and as a result, agents beliefs are filled with data. With regards to data

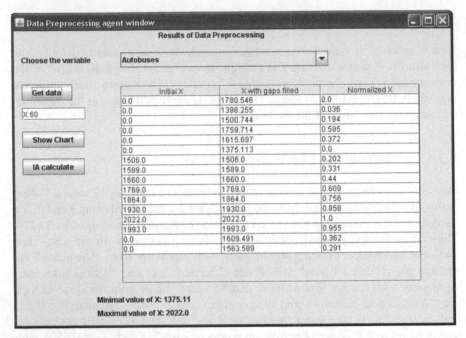

Fig. 5.4 The window with results of Data Clearing: elimination of outliers, filling gaps and normalization, where $X1$ is the data before and $X2$ is the data after filling the gaps

quality and its initial preprocessing, several data sets related to pollutants appear to have many missing values and outliers (between 40 and 87 %). Therefore, the factors, which correspond to that data sets, are eliminated from the study. The data sets that contain less number of gaps are preprocessed. First, the outliers are eliminated. Second, missing values are filled using the "gold ratio" method. Third, data sets are smoothed with exponential smoothing, and, finally, normalized. As a result of these procedures, initial information is transformed and prepared for knowledge discovery.

5.3.6 Decomposition Results

Decomposition of the studied complex system, "Environmental pollution - Human health", is carried out by means of correlation analysis (see Fig. 5.5).

As correlation between variables can impede the correct execution of data mining procedures and lead to false results, a set of non-correlated independent variables **X** for each dependent variable **Y** are created (see Table 5.4). The independent variables (pollutants) that obtain an insignificant correlation with the dependent variable (disease) are also included into the set. The mutual correlation between the variables of a model is also studied. The variables that have a correlation coefficient greater than 0.7 (see subsection 3.6.1.6) are marked for exclusion from the model. This procedure is applied for regressions, artificial neural networks, and so on.

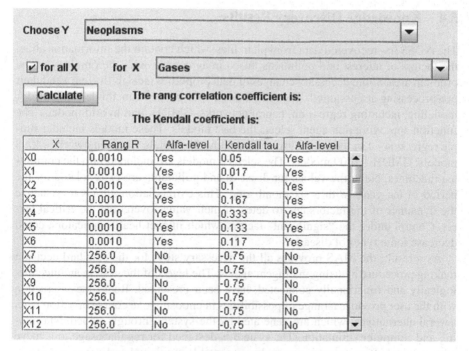

Fig. 5.5 The window that shows the decomposition

Table 5.4 Decomposition results

N.	Dependent variable	Independent variables
1	Y_1	$X_{27}, X_{35}, X_{39}, X_{40}, X_{42}, X_{54}, X_{59}, X_{60}, X_{61}, X_{62}, X_{63}, X_{64}$
2	Y_7	$X_{28}, X_{29}, X_{30}, X_{32}, X_{33}, X_{36}, X_{37}, X_{40}, X_{42}, X_{48}, X_{54}, X_{55}, X_{57}, X_{60}, X_{61}, X_{62}, X_{63}, X_{64}$
3	Y_{20}	$X_{21}, X_{24}, X_{26}, X_{27}, X_{28}, X_{29}, X_{30}, X_{31}, X_{33}, X_{35}, X_{37}, X_{38}, X_{39}, X_{40}, X_{42}, X_{44}, X_{54}, X_{55}, X_{50}, X_{60}, X_{61}, X_{62}$
4	Y_{35}	$X_8, X_9, X_{12}, X_{60}, X_{61}, X_{62}, X_{63}, X_{64}$
5	Y_{96}	$X_{592}, X_{60}, X_{61}, X_{62}, X_{63}, X_{64}$
6	Y_{100}	$X_{26}, X_{27}, X_{28}, X_{29}, X_{30}, X_{31}, X_{33}, X_{35}, X_{37}, X_{38}, X_{39}, X_{40}, X_{42}, X_{49}, X_{54}, X_{55}, X_{59}, X_{60}, X_{61}, X_{62}$
7	Y_{181}	X_6, X_{61}, X_{64}

5.4   Knowledge Discovery Results

The ADSS has recovered data from plain files, which contain the information about the factors of interest and pollutants, fused in agreement with the ontology of the problem area. Some necessary changes of data properties (scalability, etc.) and their pre-processing are assumed. The ADSS has a wide range of methods and tools for modeling, including regression, neural networks, GMDH, and hybrid models. The function approximation agent selects the best models. These models include: simple regression - 43 models; multiple regression - 24 models; neural networks - 4098 models; GMDH - 1409 models. The selected models are included into the committee machines. Next, the values for diseases and pollutants are extrapolated for the period of ten years with a six month step. This extrapolation allows to visualize the dynamics of the factors and to detect if their values overcome the critical levels. Control under the "significant" factors, which impact health indicators, could decrease some types of diseases.

As a result, the MAS provides all the necessary steps for the standard decision making procedure by using intelligent agents. The levels of the system architecture, logically and functionally connected, have been presented. Real-time interaction with the user provides a range of possibilities in choosing one course of action from several alternatives, which are generated by the system through guided data mining and computer simulation. The system is designed for regular usage to achieve adequate and effective management by responsible municipal and state government authorities. Also, both traditional data mining techniques and other hybrid and specific methods, with respect to data nature (incomplete data, short data sets, etc.) are used. The combination of different tools enable to gain quality and precision of the reached models, and, hence, to make recommendations, which are based on these models. Received dependencies of interconnections and associations between the factors and dependent variables helps to correct recommendations and avoid errors.

5.4.1   Regression Models

For every class of diseases, plotting a morbidity value against one or several pollutants, simple and multiple regressions is performed, both linear and non-linear. As a result, regression models are created of least-squared, power, exponential and hyperbolic types. Each model is evaluated with Fisher F-value. The models, which do not satisfy the F-test, are eliminated from the list of accepted models. The critical F-value for $(m-1) = 2-1 = 1$ and $2(n-1) = 2(16-1) = 30$ degrees of freedom is 4.35.

Generally, as the number of accepted regression models is low, the predictability of the best performing univariate regression models ranges from 0.48 to 0.82 for the discrimination coefficient. Fig. 5.6 and Fig. 5.7 present examples of regression models and approximation to real data.

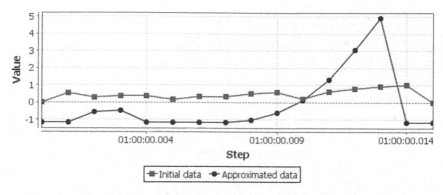

Fig. 5.6 Univariate linear regression to model $Y_0 = f(X_1)$

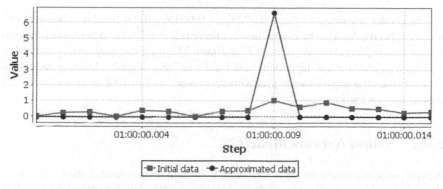

Fig. 5.7 Univariate regression to model $Y_{14} = f(X_{44})$

The model given in Fig. 5.6 is a univariate regression model, that constructs the function $Y_0 = f(X_1)$ and is equals $Y_0 = 6.42X_1 - 0.068$. The figure shows initial data, and the data approximated with the model. Visual analysis shows that the model does not fit well with the dataset. The same conclusion is drawn by analyzing its statistical criteria: the correlation coefficient $R = 0.48$, the determination coefficient $D = 0.23$ and the F-criterion $F = 4.4$. The regression model given in Fig. 5.7, which models the dependency $Y_{14} = f(X_{44})$ and has a form $Y_{14} = 4.43X_{44} - 0.144$ shows better results in fitting the initial line, as well as proving the statistical criteria: the correlation coefficient $R = 0.82$, the determination coefficient $D = 0.68$ and the F-criterion $F = 30.09$.

In general, univariate regression models for the current case study have been characterized with low values of stat indicators and have demonstrated that they cannot be used for modeling. Multiple regression models have shown better performance results. For example, the multiple regression model for the Y_{15} is given in Fig. 5.8.

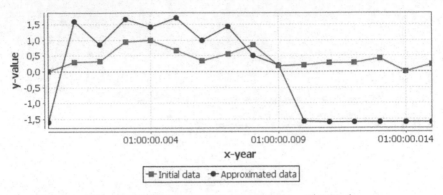

Fig. 5.8 Multiple regression to model $Y_{15} = f(X_4, X_{14})$

The model is written as $Y_{15} = 0.022X_{14} + 0.001X_4 + 0.012$ and its statistical criteria for this model are: the correlation coefficient $R = 0.77$, the determination coefficient $D = 0.59$ and the F-criterion $F = 20.69$. Meaning the explanatory variables, X_4 and X_{14}, explain the dependent variable Y_9 in 59% of cases. In other words, in 59% of cases the model would give a correct result, and in 49% of cases the model would give an incorrect result.

5.4.2 Neural Network Models

Neural network-based models, calculated for the experimental data sets, have demonstrated high performance results. Networks trained with resilient propagation and with backpropagation algorithms have similar architectures, and the training and testing procedures are equivalent. Before modeling, previous experiments are carried out. The training parameters are varied and the outputs of the neural network models are evaluated. These experiments help to determine optimal values of the parameters discussed. The best results are gotten from the networks with a limited number of hidden layers and neurons. In fact, the neural network with one hidden layered appears to be the optimal architecture for working with short data sets.

Short training sets (which indeed are used for the experiment) require networks with a simple structure. In feedforward networks trained with the backpropagation algorithm, the values of learning rate and momentum are varied within the interval [0, 0.99]. The best results are obtained with the values of the learning rate within the interval [0.85, 0.99] and the values of momentum within the range [0.3, 0.4]. Feedforward neural networks trained with the resilient propagation training algorithm (see section 3.6.2.4) obtain high performance results with the zero tolerance equal to 10^{15}, the initial update value within the range [0.05, 0.15], and the maximum step equal to 50.

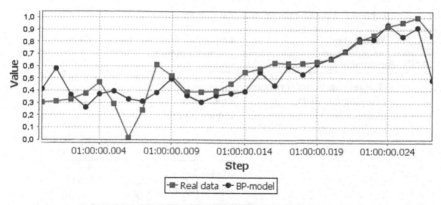

Fig. 5.9 The example of BP model for Y_{35}

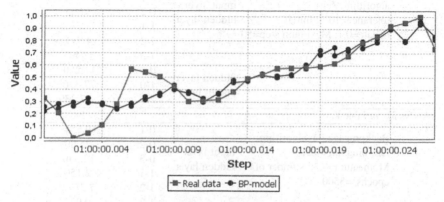

Fig. 5.10 The example of RPROP model for Y_{35}

Table 5.5 represents some of the parameters for ANN training: model type, where BP stands for "neural networks trained with backpropagation algorithm" and RPROP stands for "neural networks trained with resilient propagation algorithm". The layers include: input, hidden and output layers; activation function, though it is not given in Table 5.5, sigmoid function; values of the learning rate and momentum for BP networks and zero tolerance, initial update and maximum step for RPROP networks are given. The stopping criteria is double: the training error should not exceed 0.01 and the number of epochs should not be less than 5500.

The neural network trained with genetic algorithms uses training sets with the following training parameters:

- population size - the size of population, used for training,
- mutation percent - the percent of the population, to which the mutation operator will be applied, and,
- percent to mate - the part of the population, to which the crossover operator will be applied.

Table 5.5 Learning parameters learning for artificial neural networks

N.	Model type	Layer	Neurons	Weights
ANNs for variable Y_0				
1	Backpropagation algorithm. Learning rate=0.95; momentum=0.4; Number of epochs¿=5500	input layer - hidden layer	0-0	-1.382
			0-1	-0.195
			0-2	4.340
			1-0	12.266
			1-2	-0.246
			1-3	-0.219
		hidden layer - output layer	0-0	4.872
			1-0	0.635
			2-0	-0.176
2	Resilient propagation algorithm. Zero tolerance=10^{15}; Initial update=0.1; Maximum step=50	input layer - hidden layer	0-0	-16.326
			0-1	10.158
			0-2	-3.669
			1-1	7.230
			1-2	-4.811
			1-3	0.940
		hidden layer - output layer	0-0	11.269
			1-0	77.336
			2-0	-17.239
ANNs for variable Y_1				
3	Backpropagation algorithm. Learning rate=0.9; Momentum=0.3; Number of epochs¿=5500	input layer - hidden layer	0-1	-0.554
			0-2	-25.897
			0-3	12.140
			1-1	-2.150
			1-2	7.277
			1-2	-9.023
		hidden layer - output layer	0-0	-0.937
			1-0	7.569
			2-0	9.512
4	Resilient propagation algorithm. Zero tolerance=10^{15}; Initial update=0.1; Maximum step=50	input layer - hidden layer	0-0	-16.326
			0-1	10.158
			0-2	-3.669
			0-3	7.230
			0-4	-4.811
			0-5	0.940
		hidden layer - output layer	0-0	11.269
			0-1	77.336
			0-2	-17.239

In fact, weight optimization occurs after several populations are created, so, training error curves have step-like form. These error curves helped to set optimal parameters for GANN training. Fig. 5.11 demonstrates how this form of error training curve changes depending on values of parameters mentioned above.

Table 5.6 GMDH-models: several examples

N.	Model	R	D
1	$Y_{23} = 4.153X_{42}^2 + 1.156X_2^2 - 2.014$	**0.95**	**0.92**
2	$Y_{14} = 5.44X_{31} + 0.171X_{34} + 0.26X_{61} - 0.727$	0.67	0.44
3	$Y_{14} = 5.292X_{64}^2 + 0.161X_{60}^2 - 0.813$	**0.98**	**0.98**
4	$Y_7 = 3.704X_{28} + 0.301X_{36} + 0.29X_{38} - 1.197$	0.54	0.29
5	$Y_9 = 4.67X_{28} - 0.051X_{30} - 0.222X_{31} - 0.54X_{36} - 0.259$	0.77	0.60
6	$Y_{24} = 0.01X_1X_0 - 69.01$	0.54	0.30

A visual analysis of the curves together with received criteria results demonstrates that GANN provides better results for the given type of data series: variation of the number of iteration (900 and 1500 population created, see Fig. 5.11(a) and Fig. 5.11(b)) has influence on the value of the training error (≈ 0.30 versus ≈ 0.25). Various values for the population size used for training and mutation percent are experimented. The charts (c) and (d) at Fig. 5.11 show that the combination of a high percentage of population used for training (0.7) and for mutation (0.7) do not significantly affect the training error. However, the combination of relatively low percent of population used for training (0.3) and for mutation (0.2) has demonstrated better performance indicators for models, as the training error is reduced to the value of ≈ 0.105.

5.4.3 Models Obtained with the Group Method of Data Handling

An important feature of the iterative GMDH algorithm is its ability to identify both linear and nonlinear polynomial models using the same approach. The results of the GMDH-modeling and the best approximation models obtained are given in Table 5.6. The best results are obtained for models number 3 and 1.

In general, GMDH-based models offer high performance results and efficiency when working with short data sets. The models are obtained with combinatorial algorithm, where the combination of the following polynomials are used: X, X^2, X_1X_2, $X_1X_2^2$, $X_1^2X_2$, $X_1^2X_2^2$, $1/X$, $1/(X_1X_2)$. The selection of the model is stopped when the regulation criterion starts to reduce.

The GMDH model given in Fig. 5.12 is created for the variable Y_{23} *Disease : Mental behavior and disorders, age group : under 1 year.* and has a form $Y_{23} = 4.153X_{42}^2 + 1.156X_2^2 - 2.014$. The statistical characteristics of this model are the following: the correlation coefficient $R - 0.95$, the determination coefficient $D = 0.91$ and the F-criterion $F = 50.825$. This model fits well the experimental data and has high values of statistical parameters. The model $Y_{14} = 5.292X_{64}^2 + 0.161X_{60}^2 - 0.813$ for the variable Y_{14} *Disease: Certain conditions originating in the prenatal period. Age group: all the ages*, which is shown in Fig. 5.13, demonstrates a very high statistical performance results (see Table 5.6).

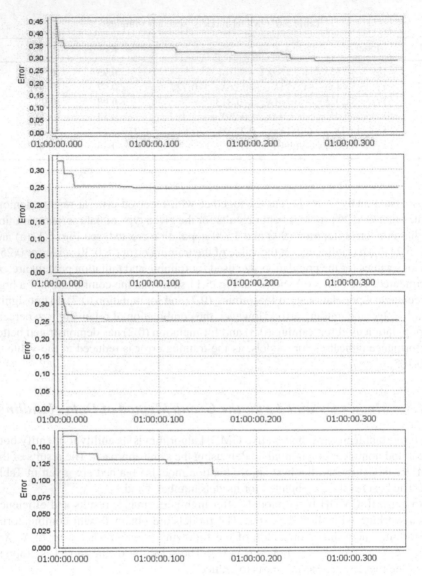

Fig. 5.11 Error functions for neural networks training with genetic algorithms with different parameters

The visual analysis shows that this model almost perfectly fits initial data. Models with similar characteristics such as the model shown in Fig. 5.13, are candidates to be included into the hybrid committee for the given variable.

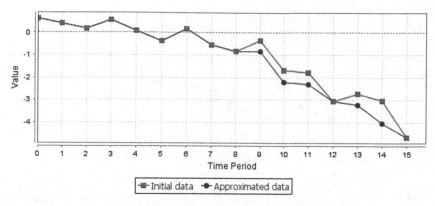

Fig. 5.12 The GMDH model for Y_{23}

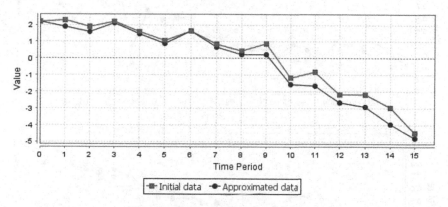

Fig. 5.13 The GMDH model for Y_{14}

5.4.4 Committee Machines

A final model for every variable is a committee machine. As an example of a committee machine, the outcomes of modeling for the variable of interest Y_{35} : *Disease: External causes of death. Age group: all the ages* are discussed. First, after the decomposition of the number of variables (pollutants) that could be included into models for interest Y_{35} is reduced and includes the following factors: X_8, X_9, X_{12}, X_{60}, X_{61}, X_{62}, X_{63}, X_{64}. Several models that include these factors are created for the variable, Y_{35}, and are then evaluated and the best are selected. These are the best models obtained:

1. Multiple regression model $Y_{35} = f_1(X_9, X_{61})$.
2. Neural network trained with backpropagation algorithm $Y_{35} = f_2(X_8, X_{63}, X_9)$ (see Fig. 5.14).

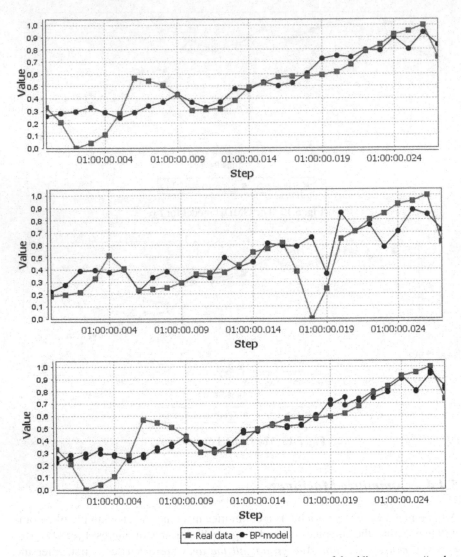

Fig. 5.14 Accepted models for the variable Y_{35} "External causes of death", age group "under 1 year". Approximation of real data by BP-trained (above) and RPROP-trained (below) neural networks

3. Neural network trained with RPROP algorithm $Y_{35} = f_3(X_{60}, X_{62}, X_{12})$ (see Fig. 5.14).
4. Neural network trained with genetic algorithms $Y_{35} = f_4(X_{64}, X_{12})$.

The final model generated by the committee machine is:

$$Y_{35} = \frac{f_1(X_9, X_{61})R_{f_1} + f_2(X_8, X_{63}, X_9)R_{f_2} + f_3(X_{60}, X_{62}, X_{12})R_{f_3} + f_4(X_{64}, X_{12})R_{f_4}}{R_{f_1} + R_{f_2} + R_{f_3} + R_{f_4}} \quad (5.1)$$

where f_i is a model, included into the committee machine, and R_{f_i} is the correlation coefficient for the $i - th$ model, $i \in [0, \ldots, n]$, where n is the number of models.

Fig. 5.14 gives graphical representation of the models. The factual information covers 28 years, which are given with a six months step. It starts at "0" and finishes at "27.5". The forecast is made for 10 years, and includes the marks starting from "28" and finishing with "37.5". To realize the forecast, the autoregressive neural networks models for all the factors from the formula of the committee machine (see 5.1) are calculated. The autoregressive model is a prediction formula that predicts an output $y(n)$ of a system based on the previous outputs, $y(n-1), y(n-2) \ldots$ and inputs, $x(n), x(n-1), x(n-2) \ldots$.

For the current case, each autoregressive model is calculated as $x(t) = f(x(t-1), x(t-2), \ldots, x(t-4))$, where t represents time, and has values $(1, 2, \ldots, n)$, n is the length of the data set, and $x(t)$ is the value of the factor at step t. Furthermore, each autoregressive neural network model belongs to the feedforward type, and is trained with RPROP algorithm. Its structure includes an input layer with five input neurons, a hidden layer with three or four neurons, and an output layer with one neuron. When the predictions for the factors from the formula of the committee machine are received, they are used to calculate the forecast for Y_{35}.

The models show similar results that do not vary much. In accordance with the forecast, the morbidity from external causes has a tendency to decline. For the period of prediction, all the models give similar forecasts, which are not strongly dispersed. That similarity in predictions by different models proves the tendency of the situation development. The results obtained by the committee machine in accordance with the equation 5.1 is composite response from the best models.

5.4.5 *Environmental Impact Assessment Results*

The impact assessment has shown the dependencies between water characteristics and neoplasm, complications of pregnancy, childbirth and congenital malformations, and deformations and chromosomal abnormalities. Table 5.7 shows the outcomes of impact assessment for several variables of interest (classes of diseases), which proves that within the most important factors apart from water pollutants, there are indicators of petroleum usage, mine output products and some types of wastes.

Fig. 5.16 to Fig. 5.19 give a view of impact assessment results for selected disease classes. Fig. 5.16 shows a pie-diagram, which demonstrates the reults of the impact assessment for the class "Diseases of the respiratory system" and for the age group "1-4 years". Corresponding with the diagram, pollutants originated by products of the mining industry are the main influencing factors. Fig. 5.17 shows the results of impact assessment for the class of the disease "Pregnancy, childbirth and the puerperium". It demonstrates that the most influencing factors for that class

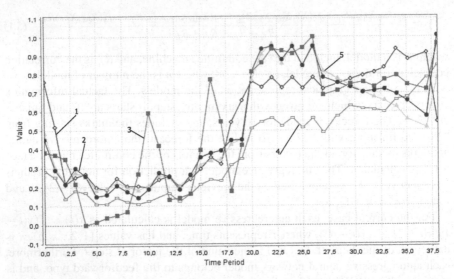

Fig. 5.15 Models for the variable Y_{35} and prognosis for the determined period. Dependent variables are X_8, X_9, X_{12}, X_{60}, X_{61}, X_{62}, X_{63} and X_{64}. The data gotten by the committee machine is marked as "1", the data obtained by the neural network trained with RPROP algorithm is marked as "2", the data calculated by the neural network trained with backprop-agation algorithm is marked as "3", the data obtained by the neural network trained with genetic algorithms is marked as "4", and the data received by the multiple regression model is marked as "5"

of diseases are the pollutants from the "Usage of petroleum products" group. Fig. 5.18 presents a pie diagram of the impact assessment for the class "Diseases of the blood and blood- forming organs and certain disorders involving the im-mune mechanism". There are many influencing factors in this diagram which in-clude water pollutants ("Solids in suspension","BOD$_5$"), wastes ("Dangerous metal-lic wastes" and "Dangerous wastes of paper industry") and pollutants originated by products of the mining industry ("Salt and Mercury miner products", "Hull", "Kaolin", "Gypsum", and others).

Fig. 5.19 shows a diagram of the evaluation of the environmental impact assess-ment for the disease class "Diseases of the nervous system". In accordance with the diagram, the strongest relation is determined with "Dangerous metallic wastes", "Dangerous common wastes" and "Other dangerous chemical wastes".

For the disease class "Neoplasms", the major influence indicator reveals for all age groups and with the pollutants belonging to the "Dangerous chemical wastes" group. For age groups "under 1 year" and "5 - 14 years", the following factors with a high level of influence have been identified: "BOD$_5$", "Asphalts" and "Dangerous wastes"; " 1 -4 years" the influencing factors are: "Petroleum liquid gases" and "Principal miner products". The age groups "15 - 24 years", "25 - 34 years" and "35 - 44 years" show a minor number of dependencies with influencing factors, apart from ones with "Principal miner products" and some types of dangerous wastes.

Table 5.7 Table with the outcomes of impact assessment for selected diseases

N.	Disease Class	Pollutant, which influence upon the disease
1	Neoplasm	Nitrites in water; Miner products; BOD_5; Asphalts; Dangerous chemical wastes; Fuel-oil; Petroleum liquid gases; Water: solids in suspension; Non-dangerous chemical wastes.
2	Diseases of the blood and blood- forming organs, the immune mechanism	BOD_5; Miner products; Fuel-oil; Nitrites in water; Dangerous wastes of paper industry; Water: solids in suspension; Dangerous metallic wastes.
3	Pregnancy, childbirth and the puerperium	Kerosene; Petroleum; Petroleum autos; Petroleum liquid gases; Gasohol; Fuel-oil; Asphalts; Water: COD; BOD_5; Solids in suspension; Water: Nitrites.
4	Certain conditions originating in the prenatal period	Non-dangerous wastes: general wastes; mineral, constriction, textile, organic, metal wastes. Dangerous oil wastes.
5	Congenital malformations, deformations and chromosomal abnormalities	Gasohol; Fuel-oil; COD in water; Producing asphalts; Petroleum; Petroleum autos; Kerosene; Petroleum liquid gases; Water: BOD_5, Nitrites, Solids in suspension.

In the elder age groups: "55 - 64 years" significant influences are detected with all the pollutants besides "Petroleum autos" and "Asphalts", outputs for the group "75 - 84 years" shows similar results, with the exceptions of "Petroleum liquid gases", "Petrol", "Kerosene" and "Fuel-oil". The groups "65 - 74 years" and "more than 85" reveal influence with "Principal miner products" and "Wastes (dangerous and non-dangerous)".

Strong relationships are found between "Dangerous animal and vegetable wastes" and "Mental and behavior disorders", between "Dangerous wastes of paper industry" and "Diseases of digestive system", and between "BOD_5", "Salt and Mercury miner products" and "Symptoms, signs and abnormal clinical and laboratory findings". The disease class "Congenital malformations, deformations and chromosomal abnormalities" has been related to water contaminants and to indicators of the usage of petroleum-containing products.

5.5 Decision Making

For this case study, a decision is to made by the specialist, however, information that could help him/her to ground it, is offered by the system. First, models in the form of committee machines and predictions are created, and hidden patterns and possible tendencies are discovered. Second, the results of impact assessment explain the

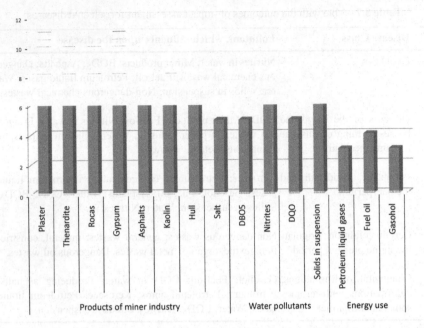

Fig. 5.16 Impact assessment diagrams for the class "Diseases of the respiratory system"

qualitative and quantitative dependencies between pollutants and diseases. Finally, the possibility of simulation is supported by the ADSS.

The variable Y_{35} "External causes of death" and the age group "under 1 year" are chosen in order to exemplify how simulation can be organized. The committee machine model for the variable of interest, Y_{35}, given in formula 5.1, is used. Suppose that there is a need to change the value of a pollutant and observe how a specific morbidity class would respond to this change. Suppose that the pollutant is X_{62}. There are five models, which compose a committee machine for the variable Y_{35}, and the "RPROP-model" includes X_{62} as an input variable. Table 5.8 shows the results of the sensitivity analysis. The second column contains values predicted by a model, the others contain values of Y_{35} calculated under the hypothesis that the variable X_{62} is going to vary. With this aim the values of the variable X_{62} are increased to 50 and 10 percents (see third and fifth columns of Table 5.8) or decreased to 50 and 10 percents (see fourth and sixth columns of Table 5.8). The model is characterized with correlation coefficient $R=0.904$ and $F=7.354$ ($F > F_{table}$). The determination coefficient, D, shows that the variables X_{62} and X_1 explain approximately 81.8% of the variable Y_{35}. The values of the variable Y_{35} are given in a normalized scale, and represent relative growth/decrease of the process.

Calculations, similar to the example presented in the Table 5.8, are made for each variable of interest. Recommendations given in Table 5.8 show possible changes in case variable X_{62} decreases or increases 10 or 50 percent.

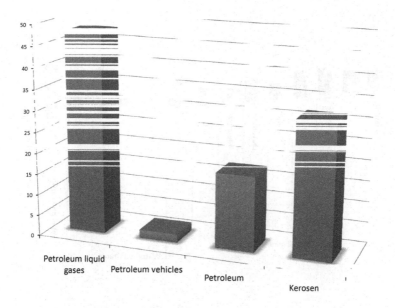

Fig. 5.17 Impact assessment diagrams for the class of the disease "Pregnancy, childbirth and the puerperium"

Table 5.8 Simulation for the variable Y_{35} : "External causes of death", age group "less than 1 year" and the dependent variable X_{62}. R=0.904, D=0.818, MAE=0.099, $MSE = 4.0E - 4$, F=7.354

Step	Predicted value	Changes of dependent variable, in %			
		+50	-50	-10	+10
1	0.537	0.583	0.401	0.474	0.510
2	0.520	0.549	0.435	0.48	0.503
3	0.520	0.55	0.434	0.48	0.504
4	0.598	0.704	0.28	0.449	0.534
5	0.614	0.737	0.247	0.443	0.541
6	0.498	0.504	0.48	0.489	0.494
7	0.605	0.72	0.264	0.446	0.537
8	0.602	0.713	0.271	0.448	0.536
9	0.515	0.54	0.444	0.482	0.501
10	0.609	0.727	0.257	0.445	0.539

Fig. 5.20 shows the window for simulation. The window has fields "Disease" and "Age" that permit the selection of the disease and the age group respectively. The "Pollutants" field permits the selection of pollutant or pollutants. The selected pollutant can be changed or modified and the forecast of the disease is calculated.

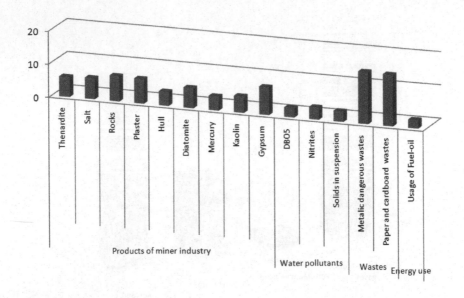

Fig. 5.18 Impact Assessment diagrams for the class "Diseases of the blood and blood- forming organs and certain disorders involving the immune mechanism"

It is possible to use default variable change limits, which include the calculation of four alternatives: independent variable (pollutant) change to +50%, -50%, +10% and -10% (see example given in Table 5.8). Approximated and forecasted values for the disease calculated by the committee machine and models are shown in Table 5.8.

Fig. 5.20 shows the table with the columns:

- "Data", which contains real data,
- "RPROP-model", which contains approximated and forecasted values calculated by the neural network trained with RPROP algorithm,
- "BP-model", which contains approximated and forecasted values calculated by the neural network trained with backpropagation algorithm,
- "GANN", which contains approximated and forecasted values calculated by the neural network trained with genetic algorithms, and,
- "Committee", which contains approximated and forecasted values calculated by the committee machine model.

It is possible to scroll through the table to view all the data. There is another table titled "Statistical criteria", which shows values of the correlation coefficient, R, determination coefficient, D, mean absolute error, MAE, mean squared error, MSE (or training error for neural-network based models), and the value of the F-criteria for each of the models. The models in Fig. 5.20 are characterized with high

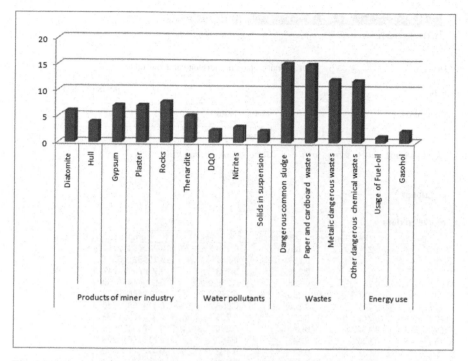

Fig. 5.19 Impact Assessment diagrams for the disease class "Diseases of the nervous system"

values, where F-criteria lies in the range from 5.785 to 7.317 and the determination coefficient, D, lies between 0.736 and 0.817.

Fig. 5.20 shows the charts for the simulation of the variable, Y_{35} : *External causes of morbidity and mortality*, in the case when the predictions are calculated with different models and the variable X_{12} is changed to +10%. Fig. 5.21 shows the outcomes of the simulation for the same variable, Y_{35}, in the case when the independent variable, X_{12}, is increased to +30%, +20% and +10%.

5.6 Discussion of the Experiment

The heterogeneous retrospective information, which formed a factual foundation for the research, was stored in CSV, DOC and XLS-files. After data retrieval, they were cleaned and preprocessed. However, some variables related to contaminants were excluded because of a high number of gaps, as originally presented. To make process modeling possible, variables were checked for correlation and inter-correlation, and divided into groups. Each group consisted of a dependent variable of interest (which represented morbidity class) and an independent variable, which represented pollutants, and did not correlate significantly with the variable of interest.

Further modeling helped to discover non-linear relationships between indicators of human health and pollutants, and generate linear and non-linear mathematical

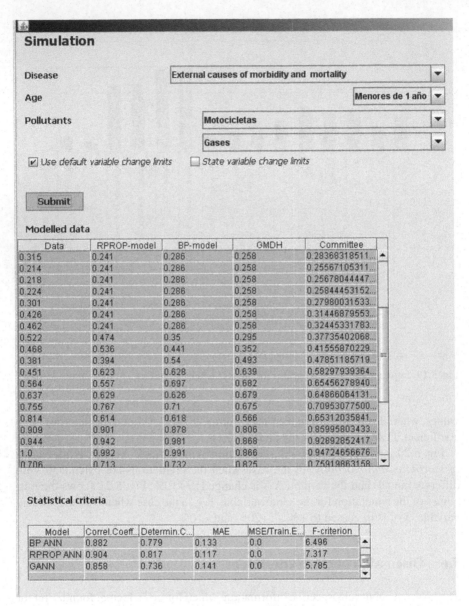

Fig. 5.20 Simulation and sensitive analysis for the variable Y_{35}

models, based on hybrid techniques, which included different types of regressions and artificial neural networks. Our results show that models, based on hybrid collective machines and neural networks could be implemented in decision support systems, as having demonstrated good performance characteristics in approximation

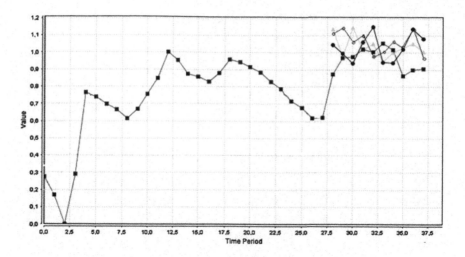

Fig. 5.21 Simulation and forecasting for the variable Y_{35}

and forecasting. This performance gap confirms a non-linear relationship between factors influencing morbidity.

Finally, different predictions of possible morbidity states under various assumptions were calculated with hybrid and neural network-based models. The predictions were generated with sensitivity analysis for the cases of explanation variables increasing and decreasing.

Chapter 6
Conclusions

Life can be only understood backwards, but can only be lived forwards.
Unknown

Abstract. This chapter contains the conclusions of this work and provides a list of activities for the near future.

6.1 Conclusions

The DeciMaS framework has been presented and discussed. The framework is destined to facilitate complex systems analysis, simulation, and to generate sets of composite decisions. DeciMas has the following stages: Preliminary domain and system analysis, System design and coding, and Simulation and Decision Making. The contents of each stage and its place within the DeciMaS organization and work flow have been explained and illustrated with examples. A three-layered architecture of the agent-based decision support system has been discussed. The meta-ontology of the domain of interest, the decision support system and their mapping have also been outlined.

The data mining methods used in the DeciMaS have also been presented, reviewed and discussed. There is a wide range of methods, including those for data preparation tasks: data clearing, missing values and outliers detection and correction, methods of normalization and integration or fusion; for knowledge discovery: statistical approaches for estimation and prediction, methods of artificial intelligence and hybrid models; for decision making: computer simulation and methods of decision theory. In addition to data mining methods, methods for model evaluation and ranging have also been introduced. The methods discussed are included into the DeciMaS framework, and the choice is based on their practical usage and efficiency, and provides support to all stages of the DeciMaS framework.

The other important characteristic of the DeciMaS framework consists in its integration possibilities: integration occurs in several different dimensions: distally, along the full length of the causal chain from remote data sources to ultimate decision alternatives; laterally, across different sources, pathways of propagation and outcomes of data mining procedures; sectorally, across different logical layers; and

M.V. Sokolova, A. Fernández-Caballero: Decision Making in Complex Systems, ISRL 30, pp. 169–172.
springerlink.com   © Springer-Verlag Berlin Heidelberg 2012

temporally over different time dimensions from analysis of retrospective backwards information to future predictions and short- to long-term forecasts. The framework also explicitly incorporates and links traditional decision support approaches with agent-oriented methodologies and knowledge discovery techniques. Furthermore, it expands to cover more complex, multi-sectoral issues and allows the user to generate administrative policies.

The monograph is dedicated to the review and analysis of existent agent-based decision support systems, and brings together existing methods, proposing an interdisciplinary flexible methodology for complex, systemic domains and policies. The motivation and objectives of the research are presented and discussed in Chapter 1. Another conclusion is the necessity to work out a new agent-based framework for decision support system creation as intelligent agents are, by nature, the best way to study a system from a multi-focal view using an interdisciplinary approach.

The review of the current state of the art, as presented in Chapter 2, introduces an overview of comprehensive approaches to complex systems study with a particular emphasis on the problem of decision making for such systems. It is outlined that the problem lies in the existence of many overlapping methodologies which intend to manage complex systems, but, nevertheless, they do not comply with the requirements for integrated support in decision making.

The DeciMas framework, described in Chapter 3, is a systematic sequence of methods that can be applied to study a complex system with respect to its systemic properties: emergency, possibility to be divided into subsystems, existence of various types of internal and external relations, etc. The approach towards a meta-ontology creation as well as its description are presented and discussed. All the evidence presented in Chapter 3 points to the fact that the DeciMaS framework appears to be a consolidated set of interdisciplinary methods and techniques which can be applied to any complex domain. This is achieved just by changing the domain ontology of the meta-ontology used in the framework.

Next, Chapter 4 provides a case study of the DeciMaS application on a domain that is characterized by high complexity and weak predictability: an "Environment - Human health" system. A meta-ontology creation with Protégé Knowledge Editor is proposed and explained in detail. In addition, a multi-agent architecture for a decision support system is shown. The sequence of the steps for the DeciMaS framework designed with the Prometheus Development Kit and its implementation with JACK Development Environment are presented as well.

In Chapter 5 the results obtained of the experiment for the selected regions are shown. The quality of the initial information is discussed and information sources that were used for data retrieval and fusion are presented. All the stages of information processing: data preprocessing: data clearing, missing values and outliers detection and correction, methods of normalization and integration or fusion; methods for knowledge discovery: statistical approaches for estimation and prediction, methods of artificial intelligence and hybrid models; decision making approaches: computer simulation methods are demonstrated and exemplified.

Taking everything into account, it is important to outline the advantages of the DeciMaS framework. First of all, it supports the general standard flow of steps for information systems' life cycle. It makes the DeciMaS framework useful and general for applications across a wide range of complex domains. Second, the possibility of the DeciMaS framework to be easily adapted to any domain of interest should be noted. The framework is organized in such a way that the change of domain is realized during the first stage of the DeciMas framework, but all the further procedures of data mining and decision generation are completed in a similar way for various domains. This characteristic adds flexibility to the DeciMaS framework and widens its application areas. Moreover, the usage of agent teams enables to distribute, to control, and to synchronize the work flows within the system. These are supervised and organized by the team leader agents to manage autonomous knowledge discovery.

Additionally, the DeciMaS framework uses known terminology, and integrates tools and methods from various disciplines, making good use of their strong sides. This facilitates the application of the DeciMaS framework by non scientific users. One of the most important insights, which can be drawn, is that there are no possibilities to elaborate a unified methodology for creation of specified and precise decision support systems for complex domains. This goal is difficult to achieve, because the commonalities shared by diverse complex systems must be taken into account without losing their specific features. However, the DeciMaS framework offers a flexible approach to the solution. Its potential should be used for complex systems study, as it helps to pool together a total capacity of methods and tools from various corresponding disciplines by means of intelligent agents.

6.2 Future Work

The future work is focused to a set of works oriented to add more data mining methods in order to widen the system's functions. These methods include classification and pattern recognition methods:

- Distance-based algorithms, which classify objects by the dissimilarity between them as measured by distance functions: Euclidean distance, city block distance, tangent distance and by the k-nearest neighbors method.
- Decision trees, that is, a predictive modeling technique for the classification of heterogeneous data, which includes methods such as ID3, C4.5, C5.0, and CART.
- Clustering, applied to retrieved data, and special clustering algorithms to large databases to scan and search for similarities. For the first task, hierarchical and partition methods can be used, while for the second one, widely used algorithms, namely BIRCH, DBSCAN, and CURE can be used.

The second objective of the future work is the implementation of special methods for decision evaluation such as the Bayesian approach, which is based on a probability model of the observed or given information. It estimates the likelihood of some property given some input data as evidence. The Bayesian approach provides a natural and flexible way to evaluate decisions and make recommendations for complex systems in situations of uncertainty and risk.

Another aim is to apply the system to other complex domains. A real-time surveillance system is an example of a possible application domain. In this case, the system would need to incorporate an additional pool of data mining techniques, dealing with image recognition, segmentation, reconstruction, etc. This application would enrich the system's storage of data mining methods.

Next, it seems necessary to apply the DeciMaS framework to complex data with multiple heterogeneity:

- represented in various formats, such as databases, texts, images, sounds and others;
- diversely structured (relational databases, XML document repositories, etc.);
- distributed and originated from several different sources, including Web;
- composites of several points of view, for example, data obtained from various types of sensors, that monitor the same scene, data expressed in different scales or languages, etc.

Solving these problems may inspire the DeciMaS framework enrichment with additional functions besides the meta-ontology creation. One of such additional functions is the accessibility of the meta-data information, which enables users to better understand the information and to exploit the warehouses. Moreover, the kit of the data mining methods may involve the use of processing tools such as crawlers and classifiers.

References

1. Anderson, B.F.: The Three Secrets of Wise Decision Making. Single Reef Press (2002)
2. Annicchiarico, R., Cortés, U., Urdiales, C.: Agent Technology and e-Health. Whitestein Series in Software Agent Technologies and Autonomic Computing. Birkhäuser, Basel (2008)
3. Aschengrau, A., Weinberg, J., Janulewicz, P., Gallagher, L., Winter, M., Vieira, V., Webster, T., Ozonoff, D.: Prenatal exposure to tetrachloroethylene-contaminated drinking water and the risk of congenital anomalies: a retrospective cohort study. Environmental Health 8(1), 814–830 (2009)
4. Athanasiadis, I.N., Mitkas, P.A.: An agent-based intelligent environmental monitoring system. Management of Environmental Quality: An International Journal 15(3), 238–249 (2004)
5. Athanasiadis, I.N., Mitkas, P.A.: Social influence and water conservation: An agent-based approach. Computing in Science and Engineering 7(1), 65–70 (2005)
6. Barchetti, U., Guido, A.L., Pulimeno, E., Bucciero, A., Mainettir, L., Sabato, S.S., Capone, L., Paiano, R.: How can ontologies support enterprise digital and paper archives? a case study. In: Proceedings of the 5th International Conference on Soft Computing as Transdisciplinary Science and Technology, CSTST 2008, pp. 627–636. ACM (2008)
7. Basson, L., Petrie, J.G.: An integrated approach for the consideration of uncertainty in decision making supported by life cycle assessment. Environmental Modelling & Software 22(2), 167–176 (2007)
8. Bauer, B., Müller, J.P., Odell, J.: Agent UML: A formalism for specifying multiagent software systems. International Journal of Software Engineering and Knowledge Engineering 11(3), 207–230 (2001)
9. Bellifemine, F.L., Poggi, A., Rimassa, G.: Developing Multi-agent Systems with JADE. In: Castelfranchi, C., Lespérance, Y. (eds.) ATAL 2000. LNCS (LNAI), vol. 1986, pp. 89 103. Springer, Heidelberg (2001)
10. Bergenti, F., Gleizes, M.P., Zambonelli, F. (eds.): Methodologies and Software Engineering for Agent Systems. The Agent-Oriented Software Engineering Handbook. Multiagent Systems, Artificial Societies, and Simulated Organizations Springer, New York Inc. (2004)
11. Bernon, C., Gleizes, M.-P., Peyruqueou, S., Picard, G.: ADELFE: A Methodology for Adaptive Multi-agent Systems Engineering. In: Petta, P., Tolksdorf, R., Zambonelli, F. (eds.) ESAW 2002. LNCS (LNAI), vol. 2577, pp. 156–169. Springer, Heidelberg (2003)

12. Bernon, C., Gleizes, M.P., Peyruqueou, S., Picard, G.: ADELFE: A methodology for adaptive multi-agent systems engineering. In: Engineering Societies in the Agents World III, pp. 70–81 (2003)
13. Berry, M.W., Browne, M. (eds.): Lecture Notes in Data Mining. World Scientific Publishing Company, Incorporated (2006)
14. Bögl, A., Schrefl, M., Pomberger, G., Weber, N.: Semantic annotation of EPC models in engineering domains to facilitate an automated identification of common modelling practices. In: Filipe, J., Cordeiro, J. (eds.), ICEIS 2008. LNBIP, vol. 19, pp. 155–171. Springer, Heidelberg (2009)
15. Bonczek, R.H., Holsapple, C.W., Whinston, A.: Foundations of Decision Support Systems. Academic Press (1981)
16. Booty, W.G., Wong, I., Lam, D., Resler, O.: A decision support system for environmental effects monitoring. Environmental Modelling & Software 24(8), 889–900 (2009)
17. Bordini, R.H., Dastani, M., Dix, J., El Fallah-Seghrouchni, A. (eds.): Multi-agent programming: Languages, platforms and applications. Multiagent Systems, Artificial Societies, and Simulated Organizations, vol. 15. Springer (2005)
18. Bossomaier, T., Jarratt, D., Anver, M.M., Scott, T., Thompson, J.: Data integration in agent based modelling. Complexity International 11, 6–18 (2005)
19. Bozdogan, H.: Statistical Data Mining & Knowledge Discovery. Chapman & Hall/CRC (2003)
20. Bozdogan, H.: Chance Discoveries in Real World Decision Making: Data-based Interaction of Human Intelligence and Artificial Intelligence. Springer (2006)
21. Brüske-Hohlfeld, I.: Environmental and occupational risk factors for lung cancer. Cancer Epidemiology: Modifiable Factors 472, 3–23 (2008)
22. Briggs, D.: A framework for integrated environmental health impact assessment of systemic risks. Environmental Health 7(1), 61–72 (2008)
23. Brisson, L., Collard, M.: An ontology driven data mining process. In: Proceedings of the Tenth International Conference on Enterprise Information Systems, ICEIS 2008, Barcelona, Spain, June 12–16, vol. AIDSS, pp. 54–61 (2008)
24. Burrafato, P., Cossentino, M.: Designing a multi-agent solution for a bookstore with the passi methodology. In: Proceedings of the Fourth International Bi-Conference Workshop on Agent-Oriented Information Systems (AOIS 2002 at CAiSE 2002), pp. 102–118 (2002)
25. Caire, G., Coulier, W., Garijo, F.J., Gomez, J., Pavón, J., Leal, F., Chainho, P., Kearney, P.E., Stark, J., Evans, R., Massonet, P.: Agent Oriented Analysis Using Message/UML. In: Wooldridge, M.J., Weiß, G., Ciancarini, P. (eds.) AOSE 2001. LNCS, vol. 2222, pp. 119–135. Springer, Heidelberg (2002)
26. Cao, M., Qiao, P.: Neural network committee-based sensitivity analysis strategy for geotechnical engineering problems. Neural Computing and Applications 17(5-6), 509–519 (2008)
27. Carlsson, C., Turban, E.: DSS: directions for the next decade. Decision Support Systems 33(2), 105–110 (2002)
28. Casals, A., Fernández-Caballero, A.: Robotics and autonomous systems in the 50th anniversary of artificial intelligence. Robotics and Autonomous Systems 55(12), 837–839 (2007)
29. Ceccaroni, L., Cortés, U., Sànchez-Marrè, M.: Ontowedss: Augmenting environmental decision-support systems with ontologies. Environmental Modelling and Software 19(9), 785–797 (2004)
30. Cernuzzi, L., Zambonelli, F.: Experiencing auml in the gaia methodology. In: International Conference on Enterprise Information Systems, pp. 283–288 (2004)

31. Chambers, R.L., Skinner, C.J. (eds.): Analysis of Survey Data. Wiley Series in Survey Methodology. John Wiley & Sons Ltd. (2003)

32. Chang, C.L.: A study of applying data mining to early intervention for developmentally-delayed children. Expert Systems with Applications 33(2), 407–412 (2007)

33. Chen, H., Bell, M.: Instrumented city database analysts using multi-agents. Transportation Research, Part C: Emerging Technologies 10(516), 419–432 (2002)

34. Clarke, B., Fokoue, E., Zhang, H.H.: Principles and Theory for Data Mining and Machine Learning. Springer Science + Business Media (2009)

35. Coleman, D., Arnold, P., Bodoff, S., Dollin, C., Gilchrist, H., Hayes, F., Jeremaes, P.: Object-Oriented Development: The Fusion Method, Prentice-Hall International edn. Prentice-Hall (1994)

36. DeLoach, S.A., Wood, M.: Developing Multiagent Systems with agentTool. In: Castelfranchi, C., Lespérance, Y. (eds.) ATAL 2000. LNCS (LNAI), vol. 1986, pp. 46–60. Springer, Heidelberg (2001)

37. Deloach, S.A.: Analysis and design using mase and agenttool. In: Midwest Artificial Intelligence and Cognitive Science Conference, MAICS 2001 (2001)

38. DeLoach, S.A., Wood, M.F., Sparkman, C.H.: Multiagent systems engineering. International Journal of Software Engineering and Knowledge Engineering 11(3), 231–258 (2001)

39. Denning, P.J., Dunham, R.: The profession of it: The core of the third-wave professional. Communications of the ACM 44(11), 21–25 (2001)

40. Dyer, R.F., Forman, E.H.: Group decision support with the analytic hierarchy process. Decision Support Systems 8(2), 99–124 (1992)

41. Eliseeva, I., Kurisheva, S., Kosteeva, T., Babaeva, I.M.B.: Econometrics. Finance and Statistics (2003)

42. Encog home page. version from 20 of noviembre (2009), http://www.heatonresearch.com/encog (accessed June 25, 2009)

43. Farlow, S.J.: Self-Organizing Methods in Modeling: GMDH-Type Algorithms. Marcel Dekker, Inc. (1984)

44. Fausett, L.V.: Fundamentals of Neural Networks: Architectures, Algorithms, and Applications. Prentice-Hall, Englewood Cliffs (1994)

45. Fernández-Caballero, A., Gascueña, J.M.: Developing Multi-Agent Systems through Integrating Prometheus, INGENIAS and ICARO-T. In: Filipe, J., Fred, A., Sharp, B. (eds.) ICAART 2009. CCIS, vol. 67, pp. 219–232. Springer, Heidelberg (2010)

46. Fernández-Caballero, A., López, M.T., Fernández, M.A., Mira, J., Delgado, A.E., López-Valles, J.M.: Accumulative Computation Method for Motion Features Extraction in Active Selective Visual Attention. In: Paletta, L., Tsotsos, J.K., Rome, E., Humphreys, G.W. (eds.) WAPCV 2004. LNCS, vol. 3368, pp. 206–215. Springer, Heidelberg (2005)

47. Fernández-Caballero, A., Vega-Riesco, J.M.: Determining heart parameters through left ventricular automatic segmentation for heart disease diagnosis. Expert Systems with Applications 36(2), 2234–2249 (2009)

48. FIPA: The Foundation for Intelligent Physical Agents (2007), http://www.fipa.org/about/index.html (accessed February 17, 2007)

49. Foster, D., McGregor, C., El-Masri, S.: A survey of agent-based intelligent decision support. In: 4th International Joint Conference on Autonomous Agents and Multiagent Systems, AAMAS 2005, July 25-29, 2005, Utrecht, The Netherlands, MAS* BIOMED 2005, pp. 104–112 (2006)

50. Fowler, M., Scott, K.: UML Distilled: Applying the Standard Object Modeling Language. Addison-Wesley Longman Ltd., Essex (1997)

51. Franklin, S., Graesser, A.: Is it an agent, or just a program?: A Taxonomy for Autonomous Agents, pp. 21–35 (1997)
52. Fujimoto, K., Nakabayashi, S.: Applying gmdh algorithm to extract rules from examples. Systems Analysis Modelling Simulation 43(10), 1311–1319 (2003)
53. Gascueña, J.M., Fernández-Caballero, A.: The INGENIAS Methodology for Advanced Surveillance Systems Modelling. In: Mira, J., Álvarez, J.R. (eds.) IWINAC 2007. LNCS, vol. 4528, pp. 541–550. Springer, Heidelberg (2007)
54. Gascueña, J.M., Fernández-Caballero, A.: Agent-Based Modeling of a Mobile Robot to Detect and Follow Humans. In: Håkansson, A., Nguyen, N.T., Hartung, R.L., Howlett, R.J., Jain, L.C. (eds.) KES-AMSTA 2009. LNCS, vol. 5559, pp. 80–89. Springer, Heidelberg (2009)
55. Gascueña, J.M., Fernández-Caballero, A.: On the use of agent technology in intelligent, multi-sensory and distributed surveillance. The Knowledge Engineering Review 26, 191–208 (2011)
56. Giorgini, P., Henderson-Sellers, B., Winikoff, M. (eds.): AOIS 2003. LNCS (LNAI), vol. 3030. Springer, Heidelberg (2004)
57. Giudici, P., Figini, S.: Applied Data Mining for Business and Industry. John Wiley & Sons (2009)
58. Giunchiglia, F., Mylopoulos, J., Perini, A.: The Tropos Software Development Methodology: Processes, Models and Diagrams. In: Giunchiglia, F., Odell, J.J., Weiss, G. (eds.) AOSE 2002. LNCS, vol. 2585, pp. 162–173. Springer, Heidelberg (2003)
59. Gohlke, J.M., Hrynkow, S.H., Portier, C.J.: Health, economy, and environment: Sustainable energy choices for a nation. Environmental Health Perspectives 116(6), A236–A237 (2008)
60. Gómez-Sanz, J.J., Botía, J., Serrano, E., Pavón, J.: Testing and Debugging of MAS Interactions with INGENIAS. In: Luck, M., Gomez-Sanz, J.J. (eds.) AOSE 2008. LNCS, vol. 5386, pp. 199–212. Springer, Heidelberg (2009)
61. Gómez-Sanz, J.J., Pavón, J.: Agent oriented software engineering with message. In: Proceedings of the Fourth International Bi-Conference Workshop on Agent-Oriented Information Systems (AOIS 2002 at CAiSE 2002), pp. 89–98 (2002)
62. Gonzalez-Velz, H., Mier, M., Arus, C., Celda, B., van Huffel, S., Lewis, P., Peet, A., Robles, M.: Agent-based distributed decision support system for brain tumour diagnosis and prognosis. In: International Conference on Multidisciplinary Information Sciences and Technologies, vol. I, pp. 288–292. Open Institute of Knowledge (2006)
63. Gorodetski, V.I., Karsaev, O., Samoilov, V., Konushy, V., Mankov, E., Malyshev, A.: Multi-agent system development kit: Mas software tool implementing gaia methodology. In: Intelligent Information Processing II, pp. 69–78 (2004)
64. Gorodetsky, V., Karsaev, O., Samoilov, V.: On-line update of situation assessment: A generic approach. International Journal of Knowledge-based and Intelligent Engineering Systems 9(4), 351–365 (2005)
65. Gorodetsky, V., Karsaev, O., Samoylov, V., Konushy, V.: Support for Analysis, Design, and Implementation Stages with MASDK. In: Luck, M., Gomez-Sanz, J.J. (eds.) AOSE 2008. LNCS, vol. 5386, pp. 272–287. Springer, Heidelberg (2009)
66. Gorodetsky, V., Karsaev, O., Samoylov, V., Serebryakov, S.: Agent-based distributed decision-making in dynamic operational environments. Intelligent Decision Technologies 3(1), 35–57 (2009)
67. Gorry, G.A., Scott Morton, M.S.: A framework for management information systems. MIT Sloan Management Review 13(1) (1971)

68. Gouveia, M., Cardoso, J.: Tourism information aggregation using an ontology based approach. In: Proceedings of the Ninth International Conference on Enterprise Information Systems, ICEIS 2007, Funchal, Madeira, Portugal, June 12-16, vol. DISI, pp. 569–572 (2007)

69. Group Method of Data Handling home page (2009), http://www.gmdh.net (accessed December 4, 2008)

70. Guarino, N., Giaretta, P.: Ontologies and knowledge bases: Towards a terminological clarification. Towards Very Large Knowledge Bases: Knowledge Building and Knowledge Sharing, 25–32 (1995)

71. Haag, S., Cummings, M., McCubbrey, D.J.: Management information systems for the information age, 3rd edn. McGraw-Hill (2002)

72. Haley, V., Talbot, T., Felton, H.: Surveillance of the short-term impact of fine particle air pollution on cardiovascular disease hospitalizations in new york state. Environmental Health 8(1), 42–52 (2009)

73. Han, J., Kamber, M.: Data Mining: Concepts and Techniques (The Morgan Kaufmann Series in Data Management Systems) 2nd edn. Morgan Kaufmann (2006)

74. Harmsen, F., Brinkkemper, S., Oei, J.L.H.: Situational method engineering for informational system project approaches. In: Proceedings of the IFIP WG8.1 Working Conference on Methods and Associated Tools for the Information Systems Life Cycle, pp. 169–194. Elsevier Science Inc. (1994)

75. Haykin, S.: Neural Networks: A Comprehensive Foundation. Macmillan, New York (1994)

76. Heine, C., Herrler, R., Kirn, S.: ADAPT@AGENT.HOSPITAL: Agent-based optimization and management of clinical processes. International Journal of Intelligent Information Technologies, IJIIT 1, 30–48 (2005)

77. Henderson-Sellers, B., Debenham, J., Tran, Q.-N.N.: Adding agent-oriented concepts derived from gaia to agent OPEN. In: Persson, A., Stirna, J. (eds.) CAiSE 2004. LNCS, vol. 3084, pp. 98–111. Springer, Heidelberg (2004)

78. Henderson-Sellers, B., Giorgini, P. (eds.): Agent-Oriented Methodologies. Idea Group Publishing (2005)

79. Herrler, R., Heine, C.: Approaches and tools to optimize and manage clinical processes. In: Application of Agents and Intelligent Information Technologies, pp. 39–65. IGI Publishing, USA (2007)

80. Holland, J.H.: Adaptation in Natural and Artificial Systems: An Introductory Analysis with Applications to Biology, Control, and Artificial Intelligence. MIT Press, Cambridge (1992)

81. Holman, I., Rounsevell, M., Shackley, S., Harrison, P., Nicholls, R., Berry, P., Audsley, E.: A regional, multi-sectoral and integrated assessment of the impacts of climate and socio-economic change in the UK. Climatic Change 71(1), 9–41 (2005)

82. Huang, C.Y., Chen, L.H., Chen, Y.L., Chang, F.M.: Evaluating the process of a genetic algorithm to improve the back-propagation network: A monte carlo study. Expert Systems with Applications 36(2), 1459–1465 (2009)

83. International Classification of Diseases (2008), http://www.who.int/classifications/icd/en/ (accessed July 19, 2008)

84. Iglesias, C.A., Garijo, M., González, J.C.: A Survey of Agent-Oriented Methodologies. In: Papadimitriou, C., Singh, M.P., Müller, J.P. (eds.) ATAL 1998. LNCS (LNAI), vol. 1555, pp. 317–330. Springer, Heidelberg (1999)

85. Iglesias, C.A., Garijo, M., Centeno-González, J., Velasco, J.R.: Analysis and Design of Multiagent Systems Using Mas-Commonkads. In: Rao, A., Singh, M.P., Wooldridge, M.J. (eds.) ATAL 1997. LNCS, vol. 1365, pp. 313–327. Springer, Heidelberg (1998)

86. Iglesias, C.A., Garijo, M., González Cristóbal, J.C., Velasco, J.R.: A methodological proposal for multiagent systems development extending common KADS. In: Proceedings of the 10th Banff Knowledge Acquisition for Knowledge-Based Systems Workshop, Banff, Canada, vol. 1, pp. 25–42 (1996)
87. Instituto de Estadistica de Castilla-La Mancha, http://www.ine.es/
88. ISO 14031:1999. Environmental management - Environmental performance - Guidelines (2008), http://www.iso.org/iso (accessed July 19, 2008)
89. Ivakhnenko, A.G., Ivakhnenko, G.A., Savchenko, E.A., Wunsch, D.: Problems of further development of gmdh algorithms: Part 2. Pattern Recognition and Image Analysis 12, 6–18 (2002)
90. Ivakhnenko, A.G., Ivakhnenko, G.A., Savchenko, E.A.: Twice-layered neural network self-organization for interpolation problems of artificial intelligence solution in the case of complete absence of information about small part of input variables. Systems Analysis Modelling Simulation 43(10), 1377–1382 (2003)
91. JACK Intelligent Agents home page (2008),
    http://www.agent-software.com/shared/home/
    (accessed July 10, 2007)
92. JACK Intelligent Agents ©, Agent Manual, Release 5.3. Agent Oriented Software Pty. Ltd (2005)
93. JACK Intelligent Agents ©, Design Tool Manual, Release 5.3. Agent Oriented Software Pty. Ltd. (2005)
94. JACK Intelligent Agents ©, Development Environment Manual, Release 5.3. Agent Oriented Software Pty. Ltd. (2005)
95. Jacobson, I., Booch, G., Rumbaugh, J.: The Unified Software Development Process. Addison-Wesley Longman Publishing Co., Inc., Boston (1999)
96. Jagannathan, G., Wright, R.N.: Privacy-preserving imputation of missing data. Data & Knowledge Engineering 65(1), 40–56 (2008)
97. Jakeman, A.J., Letcher, R.A.: Integrated assessment and modelling: Features, principles and examples for catchment management. Environmental Modelling & Software 18(6), 491–501 (2003)
98. Janssen, S., Ewert, F., Li, H., Athanasiadis, I.N., Wien, J.J.F., Thérond, O., Knapen, M.J.R., Bezlepkina, I., Alkan-Olsson, J., Rizzoli, A.E., Belhouchette, H., Svensson, M., van Ittersum, M.K.: Defining assessment projects and scenarios for policy support: Use of ontology in integrated assessment and modelling. Environmental Modelling & Software 24(12), 1491–1500 (2009)
99. Jena - A Semantic Web Framework for Java (2009),
    http://jena.sourceforge.net/, (accessed January 12, 2009)
100. Juan, T., Pearce, A., Sterling, L.: Roadmap: extending the gaia methodology for complex open systems. In: Proceedings of the First International Joint Conference on Autonomous Agents and Multiagent Systems, AAMAS 2002, pp. 3–10. ACM (2002)
101. Juan, T., Sterling, L.: A meta-model for intelligent adaptive multi-agent systems in open environments. In: Proceedings of the Second International Joint Conference on Autonomous Agents and Multiagent Systems, AAMAS 2003, pp. 1024–1025. ACM (2003)
102. Kaiser, R., Romieu, I., Medina, S., Schwartz, J., Krzyzanowski, M., Künzli, N.: Air pollution attributable postneonatal infant mortality in u.s. metropolitan areas: A risk assessment study. Environmental Health: A Global Access Science Source, 4–26 (2004)
103. Karaca, F., Anil, I., Alagha, O., Camci, F.: Traffic related pm predictor for besiktas, turkey. In: Proceedings of the 4th International ICSC Symposium on Information Technologies in Environmental Engineering, ITEE 2009, Thessaloniki, Greece, May 28-29, pp. 317–330 (2009)

104. Kebair, F., Serin, F.: Multiagent Approach for the Representation of Information in a Decision Support System. In: Euzenat, J., Domingue, J. (eds.) AIMSA 2006. LNCS (LNAI), vol. 4183, pp. 98–107. Springer, Heidelberg (2006)
105. Keen, P.: Decision support systems: The next decade. Decision Support Systems 3(3), 253–265 (1987)
106. Kern, H.: The interchange of (meta)models between MetaEdit+ and Eclipse EMF using M3-level-based bridges. In: 8th OOPSLA Workshop on Domain-Specific Modeling at OOPSLA 2008, University of Alabama at Birmingham pp. 14–19 (2008)
107. Kinny, D., Georgeff, M.P., Rao, A.S.: A Methodology and Modelling Technique for Systems of Bdi Agents. In: Perram, J., Van de Velde, W. (eds.) MAAMAW 1996. LNCS, vol. 1038, pp. 56–71. Springer, Heidelberg (1996)
108. Kruchten, P.: The Rational Unified Process: An Introduction, 2nd edn. Addison-Wesley Professional, Boston (2000)
109. Kumar, K., Welke, R.J.: Methodology engineering R: A proposal for situation-specific methodology construction, pp. 257–269 (1992), http://portal.acm.org/citation.cfm?id=133574
110. van Lamsweerde, A.: Goal-oriented requirements engineering: A guided tour. In: Proceedings of the 5th IEEE International Symposium on Requirements Engineering, RE 2001, pp. 249–263 (2001)
111. Larose, D.T.: Discovering Knowledge In Data: An Introduction To Data Mining. John Wiley & Sons (2005)
112. Last, M., Kandel, A., Bunke, H. (eds.): Data Mining in Time Series Databases. World Scientific (2004)
113. Leondes, C.T.: Fuzzy Logic and Expert Systems Applications, vol. 6. Academic Press (1997)
114. Leondes, C.T. (ed.): Algorithms and Architectures: Neural Network Systems Techniques and Applications, vol. 1. Academic Press (1998)
115. Letcher, R.A., Croke, B.F.W., Jakeman, A.J.: Integrated assessment modelling for water resource allocation and management: A generalized conceptual framework. Environmental Modelling & Software 22(5), 733–742 (2007)
116. Levin, M.S.: Composite Systems Decisions (Decision Engineering). Springer, New York, Inc. (2006)
117. Li, S.: Agentstra: An internet-based multi-agent intelligent system for strategic decision making. Expert Systems with Applications 33(3), 565–571 (2007)
118. Li, T.S., Su, C.T., Chiang, T.L.: Applying robust multi-response quality engineering for parameter selection using a novel neural-genetic algorithm. Computers in Industry 50(1), 113–122 (2003)
119. Little, R.J.A., Rubin, D.B.: Statistical Analysis with Missing Data. John Wiley & Sons, Inc. (1986)
120. Liu, L., Yu, E.: From requirements to architectural design - using goals and scenarios (2001)
121. Luck, M., Gomez-Sanz, J.J. (eds.): AOSE 2008. LNCS, vol. 5386. Springer, Heidelberg (2009)
122. Lussier, Y.A., Williams, R., Li, J., Jalan, S., Borlawsky, T., Stern, E., Kohli, I.: Partitioning knowledge bases between advanced notification and clinical decision support systems. Decision Support Systems 43(4), 1274–1286 (2007)
123. Lux, T., Matthews, W.A.: Advanced technology for environmental modelling. Environmental Modelling & Software 22(3), 279–280 (2007)
124. Ly, T., Greenhill, S., Venkatesh, S., Pearce, A.: Multiple hypothesis situation assessment. In: Proceedings of The 6th International Conference on Information Fusion, FUSION 2003, Brisbane, Australia (2003)

125. Madala, H.R., Ivakhnenko, A.G.: Inductive Learning Algorithms for Complex Systems Modeling. CRC Press, Inc., Boca Raton (1994)
126. Mahmoud, M., Liu, Y., Hartmann, H., Stewart, S., Wagener, T., Semmens, D., Stewart, R., Gupta, H., Dominguez, D., Dominguez, F., Hulse, D., Letcher, R., Rashleigh, B., Smith, C., Street, R., Ticehurst, J., Twery, M., van Delden, H., Waldick, R., White, D., Winter, L.: A formal framework for scenario development in support of environmental decision-making. Environmental Modelling & Software 24(7), 798–808 (2009)
127. Maier, M., Rechtin, E.: The art of systems architecting. CRC Press (2000)
128. Marwala, T.: Computational Intelligence for Missing Data Imputation, Estimation,and Management: Knowledge Optimization Techniques. Information Science Reference (an imprint of IGI Global), New York, USA (2009)
129. Migliore, E., Berti, G., Galassi, C., Pearce, N., Forastiere, F., Calabrese, R., Armenio, L., Biggeri, A., Bisanti, L., Bugiani, M., Cadum, E., Chellini, E., Dell'Orco, V., Giannella, G., Sestini, P., Corbo, G., Pistelli, R., Viegi, G., Ciccone, G., Group, S.C.: Respiratory symptoms in children living near busy roads and their relationship to vehicular traffic: Results of an italian multicenter study SIDRIA 2. Environmental Health 8(1), 27–42 (2009)
130. Miller, S.N., Semmens, D.J., Goodrich, D.C., Hernandez, M., Miller, R.C., Kepner, W.G., Guertin, D.P.: The automated geospatial watershed assessment tool. Environmental Modelling & Software 22(3), 365–377 (2007)
131. Mira, J., García, A.E.D., Fernández-Caballero, A., Fernández, M.A.: Knowledge modelling for the motion detection task: the algorithmic lateral inhibition method. Expert Systems with Applications 27(2), 169–185 (2004)
132. Myatt, G.J.: Making Sense of Data: A Practical Guide to Exploratory Data Analysis and Data Mining. Wiley-Interscience (2006)
133. Nastar, M., Wallman, P.: An interdisciplinary approach to resolving conflict in the water domain. In: Information Technologies in Environmental Engineering Proceedings of the 4th International ICSC Symposium Thessaloniki, Greece, pp. 411–424 (2009)
134. Nguyen, T.G., de Kok, J.L., Titus, M.J.: A new approach to testing an integrated water systems model using qualitative scenarios. Environmental Modelling & Software 22(11), 1557–1571 (2007)
135. Nikolaev, N.Y., Iba, H.: Polynomial harmonic gmdh learning networks for time series modeling. Neural Networks 16(10), 1527–1540 (2003)
136. Noy, N.F., Crubezy, M., Fergerson, R.W., Knublauch, H., Tu, S.W., Vendetti, J., Musen, M.A.: Protégé-2000: An open-source ontology-development and knowledge-acquisition environment. In: AMIA, Annual Symposium Proceedings / AMIA Symposium, pp. 953–964 (2003)
137. Orru, H., Teinemaa, E., Lai, T., Tamm, T., Kaasik, M., Kimmel, V., Kangur, K., Merisalu, E., Forsberg, B.: Health impact assessment of particulate pollution in tallinn using fine spatial resolution and modeling techniques. Environmental Health 8(1), 98–105 (2009)
138. Ossowski, S., Fernandez, A., Serrano, J.M., Perez-de-la Cruz, J.L., Belmonte, M.V., Hernandez, J.Z., Garcia-Serrano, A.M., Maseda, J.M.: Designing multiagent decision support system - the case of transportation management. In: AAMAS 2004, pp. 1470–1471. IEEE Computer Society, Washington, DC, USA (2004)
139. Ossowski, S., Hernández, J.Z., Belmonte, M.V., Maseda, J.M., Fernández, A., García-Serrano, A., Triguero Ruiz, F., Serrano, J.M., Pérez-de-la Cruz, J.L.: Multi-agent systems for decision support: A case study in the transportation management domain. Applied Artificial Intelligence 18(9-10), 779–795 (2004)
140. OWL Web Ontology Language Guide, http://www.w3.org/TR/owl-guide/

141. Padgham, L., Winikoff, M.: Developing Intelligent Agent Systems: A Practical Guide. John Wiley & Sons (2004)
142. Pahl-Wostl, C.: The implications of complexity for integrated resources management. Environmental Modelling & Software 22(5), 561–569 (2007)
143. Palit, A.K., Popovic, D.: Computational Intelligence in Time Series Forecasting: Theory and Engineering Applications(Advances in Industrial Control). Springer (2009)
144. Pallottino, S., Sechi, G.M., Zuddas, P.: A dss for water resources management under uncertainty by scenario analysis. Environmental Modelling & Software, Special Issue, 1031–1042 (2004)
145. Passuello, A., Bojarski, A., Schuhmacher, M., Jiménez, L., Nadal, M.: Evaluating long-term contamination in soils amended with sewage sludge. In: Information Technologies in Environmental Engineering, Proceedings of the 4th International ICSC Symposium, ITEE 2009, Thessaloniki, Greece, May 28-29, pp. 465–477 (2009)
146. Pavón, J., Gómez-Sanz, J.J.: Agent Oriented Software Engineering with INGENIAS. In: Mařík, V., Müller, J.P., Pěchouček, M. (eds.) CEEMAS 2003. LNCS (LNAI), vol. 2691, pp. 394–403. Springer, Heidelberg (2003)
147. Pavón, J., Sansores, C., Gómez-Sanz, J.J.: Modelling and simulation of social systems with ingenias. International Journal of Agent-Oriented Software Engineering (IJAOSE) 2(2), 196–221 (2008)
148. Perner, P. (ed.): ICDM 2009. LNCS, vol. 5633. Springer, Heidelberg (2009)
149. Power, D.: A brief history of decision support systems (2007), http://DSSResources.COM/history/dsshistory.html (accessed May 10, 2007)
150. Protecting health from climate change: global research priorities. WHO Press, World Health Organization Press (2009)
151. Protégé home page (2006), http://protege.stanford.edu/ (accessed December 01, 2006)
152. Rechtin, E.: Systems architecting of organizations: Why eagles can't swim. CRC Press (1999)
153. Ren, C., Tong, S.: Health effects of ambient air pollution - recent research development and contemporary methodological challenges. Environmental Health 7(1), 56–70 (2008)
154. Renger, M., Kolfschoten, G.L., de Vreede, G.J.: Challenges in collaborative modeling: A literature review. In: CIAO / EOMAS, LNBIP, vol. 10, pp. 61–77. Springer (2008)
155. Review of the EU Sustainable Development Strategy (EU SDS) (2006), http://register.consilium.europa.eu/pdf/en/06/st10/st10917.en06.pdf (accessed November 19, 2008)
156. Riaño, D., Sánchez-Marré, M., R.-Roda, I.: Autonomous Agents Architecture to Supervise and Control a Wastewater Treatment Plant. In: Monostori, L., Váncza, J., Ali, M. (eds.) IEA/AIE 2001. LNCS (LNAI), vol. 2070, p. 804. Springer, Heidelberg (2001)
157. Rojas, R.: Neural Networks: A Systematic Introduction. Springer, New York Inc. (1996)
158. Ronald, N., Sterling, L., Kirley, M.: Evaluating JACK Sim for Agent-Based Modelling of Pedestrians. In: IAT 2006: Proceedings of the IEEE/WIC/ACM international conference on Intelligent Agent Technology, pp. 81–87. IEEE Computer Society, Washington, DC, USA (2006)
159. Rooij, A.J.F.V., Johnson, R.P., Jain, L.C.: Neural Network Training Using Genetic Algorithms. World Scientific Publishing Co., Inc., River Edge (1996)
160. Rotmans, J.: Tools for integrated sustainability assessment: A two-track approach. Integrated Assessment 6(4), 35–57 (2006)

161. Rubin, D.L., Noy, N.F., Musen, M.A.: Protégé: A tool for managing and using terminology in radiology applications. Journal of Digital Imaging 1, 34–46 (2007)
162. Russell, S., Norvig, P.: Artificial Intelligence: A Modern Approach. Prentice-Hall
(1995)
163. Samoylov, V., Gorodetsky, V.: Ontology Issue in Multi-agent Distributed Learning. In:
Gorodetsky, V., Liu, J., Skormin, V.A. (eds.) AIS-ADM 2005. LNCS (LNAI), vol. 3505,
pp. 215–230. Springer, Heidelberg (2005)
164. Sanz, J.G., Mestras, J.P.: Curso de doctorado: Agentes inteligentes desarrollo de
sistemas multi-agente. la metodologa INGENIAS,
http://www.fdi.ucm.es/profesor/jpavon/doctorado/
desarrolloSMA.pdf
165. Saraga, D.E., Sfetsos, A., Andronopoulos, S., Chronis, A., Maggos, T., Vlachogiannis,
D., Bartzis, J.G.: An investigation of the parameters influencing the determination of
the number of particulate matter sources and their contribution to the air quality of an
indoor residential environment. Information Technologies in Environmental Engineering, Proceedings of the 4th International ICSC Symposium, ITEE 2009, Thessaloniki,
Greece, May 28-29, pp. 453–464 (2009)
166. Sarycheva, L.: Using GMDH in ecological and socio-economical monitoring problems.
Systems Analysis Modelling Simulation 43(10), 1409–1414 (2003)
167. Schafer, J.L.: Analysis of Incomplete Multivariate Data. Chapman & Hall/CRC (1997)
168. Schniederjans, M.J., Hamaker, J.L., Schniederjans, A.M.: Information Technology In-
vestment: Decision-Making Methodology. World Scientific Press (2004)
169. Scholten, H., Kassahun, A., Refsgaard, J.C., Kargas, T., Gavardinas, C., Beulens, A.J.:
A methodology to support multidisciplinary model-based water management. Environmental Modelling & Software 22(5), 743–759 (2007)
170. Sedki, A., Ouazar, D., El Mazoudi, E.: Evolving neural network using real coded
genetic algorithm for daily rainfall-runoff forecasting. Expert Systems with Applications 36(3), 4523–4527 (2009)
171. Segen, J.C.: McGraw-Hill Concise Dictionary of Modern Medicine. McGraw-Hill
Medical (2005)
172. Shim, J.P., Warkentin, M., Courtney, J.F., Power, D.J., Sharda, R., Carlsson, C.: Past,
present, and future of decision support technology. Decision Support Systems 33(2),
111–126 (2002)
173. Sivanandam, S., Deepa, S.: Introduction to Genetic Algorithms. Springer, Heidelberg
(2008)
174. Smirnov, A.V., Pashkin, M., Levashova, T., Shilov, N., Kashevnik, A.: Role-based decision mining for multiagent emergency response management. In: Autonomous Intelligent Systems: Multi-Agents and Data Mining, pp. 178–191 (2007)
175. Snedecor, G.W., Cochran, W.G.: Statistical methods, 8th edn. Iowa State University
Press (1989)
176. Sneha, S., Varshney, U.: Enabling ubiquitous patient monitoring: Model, decision protocols, opportunities and challenges. Decision Support Systems 46(3), 606–619 (2009)
177. Sokolova, M.V.: Agent-based decision support system for environmental impact assessment. Master's thesis, University of Castilla-La Mancha, High School of Engineering
Informatics of Albacete, Faculty of Computer Science Engineering (2008)
178. Sokolova, M.V., Fernández-Caballero, A.: An Agent-Based Decision Support System
for Ecological-Medical Situation Analysis. In: Mira, J., Álvarez, J.R. (eds.) IWINAC
2007. LNCS, vol. 4528, pp. 511–520. Springer, Heidelberg (2007)
179. Sokolova, M.V., Fernández-Caballero, A.: A Meta-ontological Framework for Multi-
agent Systems Design. In: Mira, J., Álvarez, J.R. (eds.) IWINAC 2007. LNCS,
vol. 4528, pp. 521–530. Springer, Heidelberg (2007)

180. Sokolova, M.V., Fernández-Caballero, A.: A multi-agent architecture for environmental impact assessment: Information fusion, data mining and decision making. In: ICEIS 2007 - Proceedings of the Ninth International Conference on Enterprise Information Systems, Funchal, Madeira, Portugal, June 12-16, 2007, vol. DISI, pp. 219–224 (2007)

181. Sokolova, M.V., Fernández-Caballero, A.: Agent-Based Decision Making through Intelligent Knowledge Discovery. In: Lovrek, I., Howlett, R.J., Jain, L.C. (eds.) KES 2008, Part III. LNCS (LNAI), vol. 5179, pp. 709–715. Springer, Heidelberg (2008)

182. Sokolova, M.V., Fernández-Caballero, A.: The Protégé-Prometheus approach to support multi-agent systems creation. In: ICEIS 2008 - Proceedings of the Tenth International Conference on Enterprise Information Systems, Barcelona, Spain, June 12-16, vol. AIDSS, pp. 442–445 (2008)

183. Sokolova, M.V., Fernández-Caballero, A.: Data mining driven decision making. In: Proceedings of the International Conference on Agents and Atificial Intelligence, ICAART 2009, pp. 220–225. INSTICC Press (2009)

184. Sokolova, M.V., Fernández-Caballero, A.: Modeling and implementing an agent-based environmental health impact decision support system. Expert Systems with Applications 36(2), 2603–2614 (2009)

185. Sokolova, M.V., Fernández-Caballero, A.: Multi-agent-based system technologies in environmental issues. In: Information Technologies in Environmental Engineering, Proceedings of the 4th International ICSC Symposium Thessaloniki, Greece, May 28-29, pp. 549– 562 (2009) (2009)

186. Sokolova, M.V., Fernández-Caballero, A., Gómez, F.J.: Agent-based interdisciplinary framework for decision making in complex systems. In: International Conference on Agents and Artificial Intelligence, ICAART 2010, January 22-24, Valencia (Spain), pp. 96–103 (2010)

187. Sokolova, M.V., Rasras, R.J., Skopin, D.: The Artificial Neural Network Based Approach for Mortality Structure Analysis. American Journal of Applied Science 2(3), 1698–1702 (2006)

188. Sprague Jr., R.H., Carlson, E.D.: Building Effective Decision Support Systems. Prentice Hall Professional Technical Reference (1982)

189. Sterling, L., Taveter, K.: The Art of Agent-Oriented Modeling. The MIT Press (2009)

190. Stieb, D., Szyszkowicz, M., Rowe, B., Leech, J.: Air pollution and emergency department visits for cardiac and respiratory conditions: a multi-city time-series analysis. Environmental Health 8(1), 75–90 (2009)

191. Symeonidis, A.L., Athanasiadis, I.N., Mitkas, P.A.: A retraining methodology for enhancing agent intelligence. Knowledge-Based Systems 20(4), 388–396 (2007)

192. Thangarajah, J., Padgham, L., Winikoff, M.: Prometheus design tool. In: Proceedings of the Fourth International Joint Conference on Autonomous Agents and Multiagent Systems, AAMAS 2005, pp. 127–128. ACM (2005)

193. Tolvanen, J.P., Rossi, M.: Metaedit+: Defining and using domain-specific modeling languages and code generators. In: OOPSLA 2003: Companion of the 18th annual ACM SIGPLAN conference on Object oriented programming, systems, languages, and applications, pp. 92–93. ACM (2003) (2003)

194. Torii, D., Ishida, T., Bonneaud, S., Drogoul, A.: Layering Social Interaction Scenarios on Environmental Simulation. In: Davidsson, P., Logan, B., Takadama, K. (eds.) MABS 2004. LNCS (LNAI), vol. 3415, pp. 78–88. Springer, Heidelberg (2005)

195. Tran, Q.N.N., Henderson-Sellers, B., Debenham, J.K.: Incorporating the elements of the mase methodology into agent open. In: ICEIS 2004, Proceedings of the 6th International Conference on Enterprise Information Systems, Porto, Portugal, April 14-17, pp. 380–388 (2004) (2004)

196. Urbani, D., Delhom, M.: Water Management Policy Selection Using a Decision Support System Based on a Multi-agent System. In: Bandini, S., Manzoni, S. (eds.) AI*IA 2005. LNCS (LNAI), vol. 3673, pp. 466–469. Springer, Heidelberg (2005)

197. Vassileva, M., Vassilev, V., Staykov, B., Genova, K., Dochev, D.: Multicriteria decision support system multioptima. In: ICEIS 2008 - Proceedings of the Tenth International Conference on Enterprise Information Systems, Barcelona, Spain, June 12-16, vol. AIDSS, pp. 276–281 (2008)

198. van Velzen, N., Segers, A.: A problem-solving environment for data assimilation in air quality modeling. Environmental & Software 25(3), 277–288 (2010)

199. Wagner, G., Taveter, K.: Towards radical agent-oriented software engineering processes based on aor modeling. In: Proceedings of the IEEE/WIC/ACM International Conference on Intelligent Agent Technology, IAT 2004, pp. 509–512 (2004)

200. Weiss, G. (ed.): Multiagent Systems: A Modern Approach to Distributed Artificial Intelligence. MIT Press, Cambridge (1999)

201. Werbos, P.J.: Beyond regression: New tools for prediction and analysis in the behavioral science. Ph.D. Thesis, Harvard University, Cambridge, MA (1974)

202. White, N., te WaterNaude, J., van der Walt, A., Ravenscroft, G., Roberts, W., Ehrlich, R.: Meteorologically estimated exposure but not distance predicts asthma symptoms in schoolchildren in the environs of a petrochemical refinery: A cross-sectional study. Environmental Health 8(1), 45–59 (2009)

203. Whitley, D., Hanson, T.: Optimizing neural networks using faster, more accurate genetic search. In: Proceedings of the Third International Conference on Genetic Algorithms, ICGA 1989, pp. 391–396. Morgan Kaufmann Publishers Inc. (1989)

204. Whitley, D., Hanson, T.: Optimizing neural networks using faster, more accurate genetic search. In: Proceedings of the Third International Conference on Genetic Algorithms, pp. 391–396. Morgan Kaufmann Publishers Inc., San Francisco (1989)

205. Wilson, R.L., Sharda, R.: Bankruptcy prediction using neural networks. Decision Support Systems 11(5), 545–557 (1994)

206. Wooldridge, M.: Intelligent agents. MIT Press, Cambridge (1999)

207. Wooldridge, M.J., Ciancarini, P.: Agent-Oriented Software Engineering: The State of the Art. In: Ciancarini, P., Wooldridge, M.J. (eds.) AOSE 2000. LNCS, vol. 1957, pp. 1–28. Springer, Heidelberg (2001)

208. Wooldridge, M., Jennings, N.R., Kinny, D.: The Gaia Methodology for Agent-Oriented Analysis and Design. Autonomous Agents and Multi-Agent Systems 3(3), 285–312 (2000)

209. Zhang, M., Ciesielski, V.: Using back propagation algorithm and genetic algorithms to train and rene neural networks for object detection. Department of Computer Science, RMIT, pp. 626–635 (1998)